# BEYOND THE BOX

# BEYOND THE BOX

A Step-by-Step Guide to Building, Marketing, and Growing Your Subscription Box Business

ANDREI REBROV

JONES MEDIA PUBLISHING

**Beyond the Box:** *A Step-by-Step Guide to Building, Marketing, and Growing Your Subscription Box Business*



Jones Media Publishing
10645 N. Tatum Blvd. Ste. 200-166
Phoenix, AZ 85028
JonesMediaPublishing.com

Disclaimer:

The author strives to be as accurate and complete as possible in the creation of this book, notwithstanding the fact that the author does not warrant or represent at any time that the contents within are accurate due to the rapidly changing nature of the Internet. While all attempts have been made to verify information provided in this publication, the Author and the Publisher assume no responsibility and are not liable for errors, omissions, or contrary interpretation of the subject matter herein. The Author and Publisher hereby disclaim any liability, loss or damage incurred as a result of the application and utilization, whether directly or indirectly, of any information, suggestion, advice, or procedure in this book. Any perceived slights of specific persons, peoples, or organizations are unintentional.

In practical advice books, like anything else in life, there are no guarantees of income made. Readers are cautioned to rely on their own judgment about their individual circumstances to act accordingly. Readers are responsible for their own actions, choices, and results. This book is not intended for use as a source of legal, business, accounting or financial advice. All readers are advised to seek the services of competent professionals in legal, business, accounting, and finance field.

Printed in the United States of America

ISBN: 978-1-948382-87-8 paperback

# CONTENTS

INTRODUCTION

# THE JOURNEY BEGINS

Eleven years ago, I had no clue how to launch a website or what the difference was between designer and niche fragrances. I also didn't know much about managing a distributed engineering team. Fast forward to today, and I'm leading a team of 40 talented engineers at Scentbird, a thriving business with over 600,000 active subscribers. We've come a long way, and I've learned a ton along the way.

It all began with a simple idea: making it less risky for people to buy perfume online. That idea evolved into making it easier for people to build a collection of fragrances. The first major milestone was achieving product-market fit, which took us about a year. It's one of the most crucial steps when starting a business. But, of course, there's a lot more to it than that, and I'm here to dive into the details.

Starting a business is never easy, and I faced my fair share of challenges. The first hurdle was figuring out how to create a website that was both user-friendly and visually appealing. After working for an enterprise company I had zero experience in building a web site for regular customers, so I spent countless nights teaching myself the basics of front-end development, website hosting and things like that.

After weeks of hard work, we finally launched Scentbird. The initial response was positive, but we still had a long way to go. Achieving product-market fit was our next big goal. This meant ensuring that our product met the needs and desires of our target audience. We spent a lot of time gathering feedback from our early customers and making necessary adjustments.

One of the key changes we made was offering more flexible subscription options. Initially, we only had one subscription plan, but we soon realized that people wanted more choices. By offering different plans, we were able to cater to a wider audience and increase our subscriber base.

As Scentbird grew, it became clear that I couldn't do it all on my own. I needed a team of talented individuals who shared my passion and vision. Building a team from scratch was another challenge, especially since I had little experience in hiring and managing people. I focused on finding individuals who were not only skilled but also aligned with our company values.

One of the best decisions I made was hiring a diverse team with varied backgrounds and expertise. This diversity brought fresh perspectives and innovative ideas to the table, helping us tackle problems in creative ways. Managing a distributed engineering team came with its own set of challenges, but we established clear communication channels and set regular check-ins to stay on track.

With a solid team in place and a growing customer base, it was time to scale up. This involved streamlining our operations, improving our supply chain, and enhancing our marketing efforts. We invested in automation tools to handle repetitive tasks, allowing our team to focus on more strategic initiatives.

Marketing played a crucial role in our growth. We experimented with various channels, including social media, influencer partnerships, and email campaigns. By analyzing the performance of each channel, we were able to allocate our resources more effectively and maximize our reach.

Throughout this journey, I've made plenty of mistakes. But each mistake was a learning opportunity that helped us improve and grow. One of the biggest lessons I learned was the importance of listening to our customers. Their feedback has been invaluable in shaping our product and service.

Another lesson was the need to stay adaptable. The business landscape is constantly changing, and being able to pivot when necessary is crucial for survival. There were times when we had to make tough decisions, such as discontinuing certain products or changing our pricing strategy. These decisions were never easy, but they were necessary for the long-term success of the business.

As I reflect on our journey so far, I'm incredibly proud of what we've achieved. But this is just the beginning. There are still so many opportunities to explore and challenges to overcome. I'm excited to continue growing Scentbird and making a positive impact on our customers' lives.

In the following chapters, I'll be sharing more about the steps we took to build, market, and grow our subscription box business from zero to more than 600,000 members. From the initial idea to scaling up, I'll provide insights and practical advice to help you on your own entrepreneurial journey. So, let's dive in and embark on this adventure together!

One of the most important lessons I've learned is the value of having a great team. Sure, there are successful solo founders out there, but entrepreneurship is hard. There will be times of desperation when you'll need physical and moral support. That's where co-founders come in. Even if you choose to go solo, you'll still need a strong team behind you.

I found my co-founders almost by accident. I attended a hackathon as an agile engineering coach, filling in for a sick colleague. There, I met Sergei, who would become my future co-founder. Three months later, he and our CEO Mariya asked me to join Scentbird as a technical founder. He saw my passion for building great engineering products and working with talented engineers. That passion still drives me today as I lead our engineering efforts and find great people to join our team. Soon after that we met our forth co-founder Rachel in New York and the team was ready to rock.

A great team is essential because even if your business model doesn't work initially, you can pivot and find something that does. Passionate people who love what they do will help you get through tough times and keep pushing forward.

When we started Scentbird, I knew we needed a team of talented, dedicated individuals. I wanted people who were not only skilled but also shared our vision. Hiring the right people is crucial. It's about finding those who can bring fresh perspectives and innovative ideas to the table.

Creating a strong team isn't just about hiring the right people. It's also about fostering a culture of collaboration and respect. At Scentbird, we've worked hard to build a supportive environment where everyone feels valued and heard. This culture has been key to our success.

Every business faces challenges, and Scentbird was no exception. In the early days, we struggled with product-market fit, supply chain issues, and marketing strategies. But our team's resilience and determination helped us overcome these hurdles. We learned to pivot when necessary and adapt to changing circumstances.

There was a time when our initial business model wasn't working as expected. Instead of giving up, we pivoted. We listened to our customers, analyzed their feedback, and made the necessary changes. This flexibility allowed us to find a business model that resonated with our audience.

One of the greatest strengths of our team is our passion. Passionate people are more resilient, creative, and driven. They don't give up easily, and they're always looking for ways to improve. This passion has been a cornerstone of Scentbird's success.

Looking ahead, I'm excited about the future. With a strong team in place, we're well-positioned to continue growing and innovating. We'll keep building on our successes, learning from our failures, and pushing the boundaries of what's possible.

In this book, I'll share more about how to build a strong team, foster a positive culture, and navigate the challenges of entrepreneurship. Whether you're just starting out or looking to grow your existing business, these insights will help you on your journey.

In the following chapters, I'll cover various aspects of building a subscription box business, including product development, marketing strategies, and customer retention. Each chapter will provide actionable tips and real-world examples to guide you along the way. Whether you're just starting out or looking to take your existing business to the next level, this book has something for everyone.

Once you have a team, the real work begins. One phrase that stuck with me is "discipline equals freedom" from the book Extreme Ownership. You need a disciplined team that can execute well. Ideas are cheap, but execution is what sets successful businesses apart.

Building a successful subscription business involves several key processes. First, you need to create a great digital experience. This means having someone on your team who knows—or can learn—how to build an excellent online presence. Then, you need people who can operate your business efficiently. This might sound boring, but it's crucial. You need a solid supply chain, finance, customer support, and marketing team. Merchants today aren't just about the products they sell but the services they offer. A great service is as important as a great product.

In my experience, success comes from working together as a cohesive unit. Improvements in one area often require cooperation from others. If you see something not working, it's your job to help fix it, not just point fingers. Extreme ownership means everyone is committed to the final result, helping each other, and ensuring that the team wins together.

To build a cohesive unit, communication is key. Regular check-ins, clear goals, and open channels for feedback are essential. Everyone on the team should know what the others are working on and how it fits into the bigger picture. This transparency fosters collaboration and ensures that everyone is aligned with the company's goals.

Encourage a culture where team members feel comfortable sharing ideas and voicing concerns. When people feel heard and valued, they're more likely to take ownership of their work. This sense of ownership leads to higher motivation and better performance.

That's a brief overview of my journey and some of the key lessons I've learned. In this book, we'll dive deeper into these topics and more. I'll share more stories, mistakes, and successes that have shaped my path. I hope my experiences can help you on your own journey, saving you time, money, and effort. Let's get started!

CHAPTER 1

# KEY ACQUISITION CHANNELS

Marketing today is a crucial part of every business. We can have an amazing product—the best in its class—but unless we can market it, no one will know about it, and no one will buy it. There are many ways to achieve marketing success today, but everyone who works in the subscription business knows the cheapest way to do marketing: create a viral post on LinkedIn, YouTube, Facebook, or another platform. If only it was that easy.

The best example here is the Dollar Shave Club's promotional viral video. It features their CEO, Michael Dubin, walking through a warehouse and bluntly explaining what Dollar Shave Club is all about and how great their product is. Many people think this was an overnight success, but the reality is that their viral video was actually video number 42. It took them 41 unsuccessful attempts to create a viral video. The 42nd video worked spectacularly well.

This example shows that to achieve great results in marketing, you have to iterate a lot. You have to try many different things.

Modern marketing works through lots of iterations. In Chapter One, we'll talk about these iterations in key acquisition channels like Meta (formerly Facebook) and Google advertising.

Let's dig deeper into the story of Dollar Shave Club. The company was founded in 2011, and from the start, they knew they needed a strong marketing strategy to stand out in the crowded shaving industry. Michael Dubin and his team believed in their product, but they also knew that belief alone wouldn't translate into sales. They needed something that would grab people's attention and get them talking.

The first 41 videos they produced didn't have the impact they hoped for. They tried different concepts, styles, and messages, but nothing seemed to click. It was frustrating, to say the least. But instead of giving up, they used each failure as a learning opportunity. They analyzed what went wrong, what could be improved, and how they could better connect with their audience.

Then came video number 42. It was simple, straightforward, and funny. Michael Dubin's humor and down-to-earth approach resonated with viewers. The video went viral almost immediately, racking up millions of views and catapulting Dollar Shave Club into the spotlight. It was shared widely on social media, covered by major news outlets, and generated a huge amount of buzz.

The success of the Dollar Shave Club video teaches us several important lessons about marketing:

Persistence Pays Off: It took 42 attempts to create a viral hit. Each failed attempt brought them closer to the one that worked.

Learn from Failures: They didn't see the first 41 videos as complete failures. Instead, they used them to learn and improve.

Authenticity Matters: The winning video was authentic and relatable. Michael Dubin's personality shone through, making the brand feel genuine and approachable.

Humor Works: A bit of humor can go a long way in making your message memorable and shareable.

In modern marketing, iteration is key. You need to be willing to experiment, fail, learn, and try again. This approach is especially important in the digital age, where trends change rapidly and what works today might not work tomorrow.

The story of Dollar Shave Club is a perfect example of how persistence, learning from failures, and staying authentic can lead to marketing success. As we move forward, keep these lessons in mind. Marketing is not a one-and-done task. It's an ongoing process that requires creativity, resilience, and a willingness to keep trying until you find what works.

In the upcoming chapters, we'll dive deeper into these concepts and provide actionable tips and real-world examples to guide you on your marketing journey.

## Leveraging Facebook for Marketing

One of the most popular acquisition channels today is the Meta family, which includes Facebook and Instagram. Let's start with Facebook. It's currently the most powerful platform for customer acquisition because it offers numerous ways to promote your products.

It all begins with creating a compelling product page. This landing page is where people learn about your product and engage with your brand. Make sure it's visually appealing, informative, and easy to navigate. Include high-quality images, detailed descriptions, and customer reviews to build trust and interest.

Understanding your audience's age and demographics is crucial because Facebook covers a broad spectrum of the population, both in the US and internationally. However, in some cases, you might need to target a more specific audience. Use Facebook's audience insights tool to gather data on your potential customers and tailor your page to meet their needs and preferences.

Once you've set up your brand page, it's time to start basic campaigns. This involves properly configuring your Facebook business profile, detailing your products, and launching initial campaigns. Fill out all the necessary information on your profile, including your business hours, contact details, and a link to your website. This helps customers find you and learn more about your offerings.

Launching your initial campaigns can be both exciting and daunting. Keep in mind that it's not cheap. You need to spend a significant amount of money upfront, typically around $10,000 to $15,000, for Facebook to learn who your target audience is. But once it does, its intelligent algorithm will help bring more customers to your site.

Targeting the right audience is essential for maximizing your return on investment. Start by creating a few different ad sets, each targeting a different segment of your audience. Test various demographics, interests, and behaviors to see which groups respond best to your ads. Facebook's targeting options are incredibly detailed, allowing you to narrow down your audience based on age, location, interests, and even online behavior.

Your ads need to grab attention and encourage action. Use eye-catching images or videos, clear and concise copy, and a strong call to action. Highlight the benefits of your product and why it's better than the competition. Remember, you only have a few seconds to capture someone's attention as they scroll through their feed, so make every word and image count.

Once your campaigns are up and running, monitor their performance closely. Facebook provides detailed analytics that show how your ads are performing, including metrics like click-through rates, conversion rates, and cost per acquisition. Use this data to make informed decisions about which ads to continue, which to tweak, and which to stop.

Remember, you must keep trying new things to see what works best. Experiment with different ad formats, such as carousel ads, video ads, and slideshow ads. Try different offers and promotions to see what resonates most with your audience. The more you experiment, the more you'll learn about what works and what doesn't.

Budgeting is a crucial part of your Facebook marketing strategy. While the initial investment might seem steep, it's important to view it as a long-term investment in your business. As Facebook learns more about your audience and optimizes your campaigns, you'll start to see a higher return on your investment. Be prepared to adjust your budget based on the performance of your ads and the needs of your business.

Facebook (Meta) offers a variety of tools to help you manage and optimize your campaigns. The Facebook Pixel, for example, tracks user behavior on your website and provides valuable insights into how your ads are performing. Custom Audiences and Lookalike Audiences allow you to target users who are similar to your existing customers, increasing the likelihood of conversions.

Leveraging Facebook for marketing is a powerful way to grow your business. By creating a compelling product page, targeting the right audience, crafting effective ads, and continuously iterating on your campaigns, you can maximize your return on investment and drive significant growth.

We've covered the basics of getting started with Facebook marketing. As we move forward, we'll dive deeper into advanced strategies and techniques to help you get the most out of your marketing efforts. Whether you're just starting out or looking to take your existing campaigns to the next level, these insights will provide you with the tools and knowledge you need to succeed.

Although Instagram is part of the Meta family and can be considered Facebook's younger sibling, it's a powerful advertising platform in its own right. Instagram's strength lies in its visual nature. It's all about images and videos, making it an excellent platform for showcasing products. Whether it's demonstrating how a functional product works or simply displaying its aesthetic appeal, Instagram is the place to be.

Instagram's visual nature is its greatest strength. High-quality images and videos can capture attention and engage users in ways that text alone cannot. When showcasing products, focus on clear, attractive visuals that highlight the best features. Think about the story you want to tell with your images and videos. Are you demonstrating how a product works? Or are you highlighting its aesthetic appeal? Either way, your visuals should be compelling and professional.

## Getting Started with Instagram

To start, create a business profile on Instagram. Fill out your profile completely, including a clear bio, profile picture, and contact information. Make sure your profile reflects your brand's personality and values.

Begin posting high-quality content regularly. Consistency is key. Use a mix of photos, videos, and carousel posts to keep your feed dynamic and interesting. Utilize Instagram Stories to share behind-the-scenes content, product demos, and customer testimonials. Stories are a great way to engage with your audience on a more personal level.

Engagement is crucial on Instagram. Respond to comments, answer questions, and interact with your followers. Use Instagram's features like polls, questions, and quizzes in Stories to engage your audience and encourage interaction. The more you engage with your followers, the more likely they are to become loyal customers.

One significant aspect of Instagram is influencer marketing. Influencers can amplify your reach and credibility. Partnering with influencers who align with your brand can help you reach a larger, more targeted audience. Influencers have built trust with their followers, so their recommendations can carry a lot of weight.

Start by identifying influencers who resonate with your brand and audience. Look for those with high engagement rates and authentic connections with their followers. Reach out to them with a clear proposal that outlines how you can collaborate. Whether it's sponsored posts, product reviews, or giveaways, influencer marketing can be a powerful tool to boost your brand's visibility and credibility.

Instagram Reels are another fantastic way to engage your audience. Reels are short, entertaining videos that can reach a broad audience through the Explore page. Use Reels to showcase your products in fun, creative ways. This could include quick tutorials, behind-the-scenes looks, or customer testimonials. The key is to keep it engaging and authentic.

Instagram isn't just about posting images; it's also about conversions. There are multiple ways to drive sales. You can leave a link in your profile, offer discounts and coupons in post descriptions, or use services like Linktree to provide multiple destinations with a single click.

Another increasingly popular method is converting users directly through Instagram's shopping features. Setting up Instagram Shopping allows users to browse and purchase products directly from your Instagram profile. To get started, you need to integrate your Shopify store (or another e-commerce platform) with your Instagram account. Once set up, you can tag products in your posts and Stories, making it easy for users to shop directly from your content.

A strong call to action (CTA) is essential for driving conversions. Whether you're encouraging users to visit your website, use a discount code, or check out a new product, your CTA should be clear and compelling. Use action-oriented language and create a sense of urgency to encourage immediate action.

To measure the success of your Instagram marketing efforts, use Instagram's built-in analytics. Insights provide valuable data on your audience demographics, post engagement, and overall account performance. Use this data to refine your strategy, understand what's working, and identify areas for improvement.

Instagram is a powerful tool for marketing, with its visual appeal and robust features for engagement and sales. By creating high-quality content, engaging with your audience, leveraging influencer partnerships, and utilizing Instagram's shopping features, you can maximize your reach and drive significant growth.

## Google Advertising

Google is the original behemoth of online advertising. It was one of the first large platforms where people could advertise, and it remains incredibly powerful. Let's start with sponsored search. Whenever you search for something on Google, the first few links are usually ads from brands relevant to your search. This is often the first touchpoint for potential customers.

Sponsored search ads are those top-of-the-page results that appear when you search for something on Google. These ads are a great way to capture potential customers right when they're looking for products or services like yours. Setting up sponsored search ads involves bidding on keywords that your target audience is likely to use. The more relevant and competitive your keywords, the better your ads will perform.

Google also excels at promoting products across other websites. When you visit a media site and see banners around the articles you're reading, many of these are powered by Google. These display ads can be highly effective in reaching a broader audience. They appear on millions of websites and apps within the Google Display Network, allowing you to target users as they browse their favorite sites, watch videos on YouTube, check their Gmail, or use mobile apps.

Setting up your Google ads involves several steps. First, you need to configure your Google Ads account. This includes setting up billing information and linking your Google Analytics account. Next, define your target audience based on demographics, interests, and online behavior. Connecting Google to your product feed is also crucial. This allows Google to pull your product information and display it in relevant searches and ads.

Google Ads offers advanced targeting options to ensure your ads reach the right people. You can target based on keywords, demographics, interests, geographic locations, and even the type of device being used. Bidding strategies are also essential. You can choose from options like cost-per-click (CPC), cost-per-thousand-impressions (CPM), or cost-per-acquisition (CPA). Each strategy has its benefits, and the right choice depends on your specific goals and budget.

An important aspect of Google advertising is using Google Analytics. Properly configuring Google Analytics helps you understand your marketing efforts better. You'll gain insights into which channels and campaigns convert the best, enabling you to optimize your strategies effectively. With Google Analytics, you can track metrics like user behavior, conversion rates, and ROI. This data is invaluable for making informed decisions and improving your ad performance.

Like Facebook, Google requires time to learn who converts and who doesn't. Initially, you might not see the results you hoped for, but don't get discouraged. Google's algorithm needs time to gather data and optimize your campaigns. During this learning phase, monitor your ad performance closely. Look at metrics such as click-through rates, conversion rates, and cost per conversion. Use this data to tweak your keywords, ad copy, and targeting settings.

One powerful feature of Google Ads is retargeting. Retargeting allows you to show ads to users who have previously visited your website or interacted with your brand. This is a great way to re-engage potential customers who showed interest but didn't convert. By keeping your brand top-of-mind, retargeting can significantly increase your chances of converting these users into paying customers.

A/B testing, or split testing, is another crucial part of optimizing your Google Ads. Test different versions of your ads to see which ones perform better. This could involve changing the ad copy, images, call-to-action, or even the landing page. The goal is to identify what resonates most with your audience and refine your ads accordingly.

Google advertising is a powerful tool for driving traffic and conversions. By leveraging sponsored search ads, display ads, and retargeting campaigns, you can reach a wide audience and maximize your marketing efforts. Remember to use Google Analytics to track your performance and continuously optimize your campaigns.

CHAPTER 2

# CRM MARKETING

CRM marketing is a game-changer in today's digital age. It's all about building and maintaining relationships with your customers. When you use a CRM platform, you're not just sending out random emails or texts. You're creating personalized, targeted communication that makes your customers feel valued and understood.

But let's talk about the elephant in the room: spam. The dreaded "s" word. Spam emails are the bane of everyone's digital existence. They clog up our inboxes, waste our time, and are just plain annoying. So, how did we get here? How did email marketing, which has so much potential, end up associated with spam?

This reminds me of a Monty Python sketch. In this sketch, a restaurant offers a menu where almost every dish includes spam. The repetitive, unwanted inclusion of spam in every dish became a metaphor for those unsolicited, unwanted emails we now call spam. It's kind of funny when you think about it—a comedy sketch from the 1970s shaping the way we talk about email today.

But, as email marketing grew, so did the misuse of it. People realized they could reach thousands, even millions, of people with a single click. And they did—over and over again. Unsolicited emails started flooding inboxes, and people got fed up. That's when email service providers like Google and Yahoo stepped in. They developed spam filters to catch these unwanted emails before they reached the user's inbox. These filters are pretty sophisticated, looking at things like the sender's reputation, the content of the email, and even the recipient's interaction with previous emails from that sender.

So, if you're an email marketer, how do you avoid the spam folder? How do you make sure your emails get seen and read? Well, it's all about best practices and understanding how spam filters work. First off, always get permission before sending emails. This means using opt-in methods where users explicitly agree to receive emails from you. It's not just good manners; it's the law in many places.

Next, make sure your content is relevant and valuable to your audience. If people find your emails useful, they're more likely to open them and less likely to mark them as spam. Personalization is key here. Use your CRM to segment your audience and tailor your messages to their specific needs and interests. A generic email blast is more likely to be ignored or marked as spam than a personalized message that speaks directly to the recipient.

Another tip is to keep your email list clean. Regularly remove inactive subscribers—people who haven't opened or clicked on your emails in a long time. This not only helps with deliverability but also ensures you're not wasting resources on people who aren't interested.

Subject lines are another crucial element. They need to be clear and compelling, giving recipients a reason to open your email. Avoid using all caps or excessive punctuation, as these are red flags for spam filters. And, of course, make sure your email content is high-quality. This means no misleading information, no excessive links or images, and definitely no spammy phrases like “Act now!” or “Click here!”

Timing also plays a role in effective email marketing. Think about when your audience is most likely to check their emails and try to send your messages during those times. This increases the chances of your email being seen and read.

Finally, always provide a clear and easy way for recipients to unsubscribe. It might seem counterintuitive, but making it easy for people to opt-out shows respect for their preferences and can actually improve your sender reputation.

CRM marketing is a powerful tool when used correctly. By understanding the origins of spam and learning how to avoid the common pitfalls, you can create effective email campaigns that engage your audience and drive results. Remember, the goal is to build relationships, not just send messages. When you focus on providing value and respecting your audience, you’ll find that email marketing can be one of the most effective channels in your CRM strategy.

Crafting an effective email campaign really is both an art and a science. It’s not just about throwing some words together and hoping for the best. You’ve got to start by building a solid email list. Offer something valuable to get people to sign up—maybe a discount, a free e-book, or some exclusive content. And, of course, make sure your sign-up process is easy to follow and clearly explains what subscribers will get from you.

Now, let’s talk about the actual email content. Keep it concise and engaging. Start with a compelling subject line that grabs attention.

If your subject line is boring, your email's probably going straight to the trash. Once you've got their attention, make sure the body of your email delivers on the promise of the subject line. Use a friendly, conversational tone—like you're talking to a friend. And don't forget those clear calls-to-action (CTAs). Guide your readers on what to do next, whether it's clicking a link, making a purchase, or signing up for something.

But let's dive deeper into these points.

Building a quality email list is the foundation of any successful email campaign. It's not enough to just have a lot of email addresses; you need the right ones. Start by offering something of value in exchange for sign-ups. For example, if you're running an online store, a discount on their first purchase can be a great incentive. If you're a content creator, offering a free e-book or exclusive content can attract subscribers.

Make sure your sign-up form is easy to find and simple to fill out. Nobody wants to jump through hoops just to get on your email list. And be clear about what they're signing up for. Tell them how often they'll hear from you and what kind of content they'll receive. Transparency builds trust and sets the right expectations from the start.

## Segmenting Your List

Once you have a solid list, it's time to segment it. This might sound like extra work, but it's worth it. Segmentation allows you to tailor your messages to different groups of people. For example, you can segment based on how customers have interacted with your brand. Are they new subscribers, or have they made multiple purchases? You can also segment based on preferences they've expressed or demographics like age and location.

By sending more targeted emails, you increase the relevance of your messages. And when emails are relevant, people are more likely to open them, read them, and take action. It's a win-win.

## Crafting Your Content

Now, let's get into the nitty-gritty of writing your emails. First off, your subject line. This is the first thing people see, so make it count. Keep it short and sweet, and give them a reason to open your email. Maybe it's a limited-time offer, a piece of exciting news, or something else they won't want to miss.

Once you've got their attention, don't waste it. The body of your email should be engaging and to the point. Use a conversational tone—like you're chatting with a friend. This makes your emails more relatable and enjoyable to read. And keep it concise. People are busy and their inboxes are full. Get to the point quickly and clearly.

Include clear calls-to-action (CTAs) in your emails. What do you want your readers to do next? Whether it's clicking a link, making a purchase, or signing up for an event, make it easy for them to take that next step. And don't be afraid to use buttons for your CTAs—they're more noticeable than text links and can increase click-through rates.

Finally, don't forget to test and analyze your emails. Try out different subject lines, email designs, and CTAs to see what works best. Most email marketing platforms offer A/B testing features, so take advantage of them. And once your emails are out there, track your results. Look at open rates, click-through rates, and conversion rates to see how your emails are performing. Use this data to continually improve your campaigns.

Crafting effective email campaigns takes some effort, but it's well worth it. By building a quality email list, segmenting it, creating engaging content, and testing and analyzing your results, you can create email campaigns that not only get noticed but also drive action. So, roll up your sleeves and start crafting those emails!

Avoiding the spam folder requires a combination of best practices and attention to detail. Let's dive deeper into each of these tips to understand why they're important and how to implement them effectively.

Imagine receiving an email from an address you don't recognize. Chances are, you wouldn't open it, right? That's why it's so important to use a recognizable sender name and email address. Consistency builds trust. If your subscribers know what to expect and recognize your name or brand, they're more likely to open your emails. So, choose a sender name that reflects your brand or use your personal name if you're the main face of your business. Stick to this choice across all your emails to build a sense of familiarity and trust.

Spam filters are designed to protect users from unwanted emails, and they look for certain red flags to identify spam. Words like "free," "guarantee," and "urgent" are common in spam emails, so using them can trigger the filters. All caps and excessive punctuation are also big no-nos. Instead, focus on creating subject lines and email content that are clear, honest, and relevant to your audience. For example, instead of saying "FREE OFFER! Act Now!" try something like "Special Offer Just for You." It's less likely to trigger spam filters and feels more genuine.

A clean email list is essential for good email deliverability. Over time, some subscribers will become inactive—they might change email addresses, lose interest, or simply stop engaging with your content. Continuing to send emails to these inactive subscribers can hurt your sender reputation. Email service providers track engagement, and if they see low engagement rates, they might start sending your emails to the spam folder. To avoid this, regularly clean your list by removing inactive subscribers. You can use tools to identify these users, or you can send a re-engagement campaign to try to win them back before removing them.

Email authentication protocols like SPF, DKIM, and DMARC are technical steps that help verify your emails are legitimate. SPF (Sender Policy Framework) allows the owner of a domain to specify which mail servers are allowed to send emails on behalf of that domain. DKIM (DomainKeys Identified Mail) adds a digital signature to your emails that can be verified by the receiving mail server. DMARC (Domain-based Message Authentication, Reporting & Conformance) tells receiving servers what to do if an email fails SPF or DKIM checks. These protocols help protect your domain from being spoofed and increase the chances of your emails being delivered to the inbox.

Making it easy to unsubscribe is not just a legal requirement, but it also shows respect for your subscribers' preferences. If people can't find a way to unsubscribe easily, they might mark your email as spam out of frustration. This can hurt your sender reputation and increase the chances of your emails being filtered out as spam in the future. Include a clear, visible unsubscribe link in every email. This keeps your list healthy by ensuring that only those who want to hear from you are on it, and it builds trust with your audience.

Avoiding the spam folder is all about following best practices and paying attention to the details. By using a recognizable sender name, avoiding spam trigger words, maintaining a clean email list, authenticating your emails, and providing an easy unsubscribe option, you can improve your email deliverability and ensure your messages reach your audience. So, take these steps seriously, and you'll see better results from your email marketing campaigns.

Measuring the performance of your email campaigns is crucial for continuous improvement. You need to track key metrics like open rates, click-through rates, conversion rates, and unsubscribe rates. These metrics provide insights into what's working and what's not.

Measuring how well your email campaigns are doing is super important for making continuous improvements. You've got to track key metrics like open rates, click-through rates, conversion rates, and unsubscribe rates. These numbers will tell you what's working and what isn't.

Let's break it down. Open rates show how many people are actually opening your emails. If your open rates are low, your subject lines might need work. Click-through rates indicate how many people are clicking on the links in your emails. If these rates are low, maybe your content or CTAs aren't compelling enough. Conversion rates show how many people are taking the desired action, like making a purchase or signing up for a webinar. And unsubscribe rates tell you if people are opting out of your emails—if this number is high, it's time to rethink your strategy.

A/B testing is a valuable tool for optimization. It lets you test different versions of your emails to see what works best. Try out different subject lines, email content, CTAs, and send times. For example, you can send one version of your email to half of your list and another version to the other half. Then, compare the results to see which version performed better.

Use the data from these tests to refine your strategies. If you find that emails sent in the morning get higher open rates, adjust your send times accordingly. If a certain CTA gets more clicks, use similar language in future emails. The goal is to learn from each test and continuously improve your campaigns.

## The Future of Email Marketing

As technology evolves, so does email marketing. AI and machine learning are playing an increasingly significant role in personalizing and optimizing email campaigns. Predictive analytics can help you anticipate customer behavior and tailor your emails to meet their needs.

The future of email marketing looks pretty exciting. As technology evolves, email marketing is evolving too. AI and machine learning are becoming more important in personalizing and optimizing email campaigns. With predictive analytics, you can anticipate customer behavior and tailor your emails to meet their needs.

Imagine being able to predict what your customers want before they even know it themselves. That's the power of predictive analytics. It uses data to forecast future behavior, helping you send more relevant and timely emails. For example, if a customer has been browsing a certain product category on your website, you can send them an email with personalized product recommendations. This kind of personalization can significantly boost engagement and conversions.

Interactive emails are also becoming more popular. These are emails that go beyond static content and include elements like surveys, quizzes, and personalized product recommendations. Interactive emails are more engaging and can drive higher interaction rates.

Think about it—wouldn't you be more likely to engage with an email that includes a fun quiz or a survey asking for your opinion?

These elements make the email experience more dynamic and interactive. Plus, they provide valuable data that you can use to further personalize your marketing efforts.

Ultimately, the key to successful email marketing lies in building strong relationships with your customers. Provide value, listen to their feedback, and continuously strive to improve their experience. By doing so, you'll not only avoid the spam folder but also foster loyalty and drive long-term success.

Measuring and optimizing your email performance is crucial for continuous improvement. Track key metrics, conduct A/B tests, and use the insights to refine your strategies. Embrace new technologies like AI and interactive emails to stay ahead of the curve. Most importantly, focus on building strong relationships with your customers. Provide value, listen to their feedback, and always look for ways to enhance their experience. By doing so, you'll not only avoid the spam folder but also build loyalty and achieve long-term success.

Images and graphics can enhance your emails, but don't overdo it. Too many images can slow down loading times and might get your email flagged as spam. Balance your visuals with text and ensure that your email still makes sense if the images don't load.

Another critical aspect of CRM marketing is timing. Sending your emails at the right time can significantly impact their effectiveness. Test different times and days to see when your audience is most responsive. Most studies suggest that weekdays, particularly Tuesdays and Thursdays, are the best days for email marketing, but this can vary depending on your audience.

Let's not forget the importance of mobile optimization. More people are checking their emails on their smartphones, so your emails need to look good and be easy to navigate on smaller screens. Use a responsive design that adjusts to different screen sizes and test your emails on various devices before sending them out.

Lastly, always include an easy way for recipients to unsubscribe. This might seem counterintuitive, but making it simple for people to opt out of your emails shows that you respect their preferences and helps maintain a positive reputation for your brand. Plus, keeping your list clean by removing uninterested subscribers can improve your overall engagement rates.

Effective CRM marketing through email is about respecting your audience, providing value, and staying mindful of best practices to avoid the spam folder. By personalizing your content, crafting engaging subject lines, and optimizing for both timing and mobile devices, you can build strong relationships with your customers and drive your business forward. Remember, CRM marketing is not just about sending emails; it's about creating meaningful connections that last.

To get started with CRM marketing, the first thing you need is a CRM tool. There are tons of options out there, and honestly, most of them are pretty similar. When choosing one, think about the price and whether the interface is user-friendly. If the interface feels confusing or clunky, skip it and try another. Some popular options include MailChimp and SendGrid. These are great starting points, and you can always upgrade as your needs grow.

Once you've picked a tool, it's time to configure it so your emails don't end up in the spam folder. This part might sound a bit technical, but it's crucial. You'll need to set up three key digital signatures: SPF, DKIM, and DMARC. These signatures help verify that you're the legitimate sender of your emails and that you own the domain from which they are sent.

Let's break this down step by step. When you create an account with a CRM service, head over to the settings section. There, you'll find the SPF, DKIM, and DMARC records. Follow the instructions provided by your CRM service to create these records in your domain management panel. It might take about 15 minutes for everything to be set up, but once you're done, you'll be in a much better position to ensure your emails reach your audience's inboxes instead of their spam folders.

Now, let's dive into each of these signatures a bit more. SPF is a record that tells email providers which servers are allowed to send emails on behalf of your domain. Setting up SPF helps prevent spammers from sending emails that appear to come from your domain. DKIM adds a digital signature to your emails, verifying that the email content hasn't been altered in transit and confirming it was indeed sent by you. Lastly, DMARC ties everything together, telling email providers how to handle emails that fail SPF or DKIM checks and providing you with valuable reports on email authentication.

Choosing the right CRM tool and configuring it properly might seem like a lot of work initially, but it's worth it. A well-configured CRM tool not only helps you avoid the spam folder but also boosts your credibility and ensures your messages are delivered successfully.

When you're selecting your CRM tool, don't forget to explore the additional features each platform offers. Some CRM tools come with advanced analytics, A/B testing, and automation capabilities that can take your email marketing to the next level. For instance, MailChimp offers detailed analytics and easy-to-use automation, while SendGrid is known for its robust delivery infrastructure and scalability. Depending on your business needs, these extra features can be incredibly valuable.

Moreover, integrating your CRM tool with other software you use can streamline your workflow. Many CRM platforms offer integrations with popular tools like Shopify, Salesforce, and Google Analytics. These integrations allow you to sync data across different platforms, giving you a more comprehensive view of your customer interactions and marketing performance.

Once your CRM tool is set up, it's time to start building your email list. Make sure to collect email addresses ethically by using opt-in forms on your website, offering valuable content in exchange for email subscriptions, and ensuring compliance with regulations like GDPR and CAN-SPAM. A quality email list is the foundation of successful CRM marketing.

As you begin your email marketing campaigns, keep track of key performance metrics like open rates, click-through rates, and conversion rates. These metrics will help you understand what's working and what's not, allowing you to fine-tune your strategy for better results. Regularly review your campaigns and experiment with different subject lines, email content, and send times to see what resonates best with your audience.

Choosing and configuring your CRM tool is a critical step in successful CRM marketing. By selecting a user-friendly platform, setting up essential digital signatures, and leveraging additional features and integrations, you can create effective email campaigns that engage your audience and drive results. Remember, CRM marketing is about building relationships and providing value to your customers, so always focus on delivering relevant, high-quality content that meets their needs.

## Crafting Effective Mass Emails

In the world of email marketing, there are two main types of campaigns: mass emails and trigger campaigns. Let's dive into mass emails first. These are identical emails sent to a large audience all at once. You might use mass emails to announce a new product collection, highlight new website features, or promote holiday discounts, like a Mother's Day sale.

To make mass emails effective, you need to segment your audience. Sure, you're sending the same email to many people, but segmentation boosts your chances of engagement. For example, segment by location to reach people at the right time. If you're targeting folks in New York, send your emails around 9 a.m. But if you're also aiming at California, adjust the timing for the time difference.

Other segmentation criteria can include how you acquired the email addresses—whether through your website or a giveaway—the recipients' status on your platform (like whether they are subscribers or not), and more. Once you've segmented your audience, focus on the quality of your email. Write compelling copy, check for broken links, and use attractive, appropriately sized images. Services like emailtobox.com can help ensure your emails look great.

Now, let's break down the steps to craft an effective mass email.

First, start with a clear and engaging subject line. This is the first thing your audience will see, and it needs to grab their attention. Keep it short and to the point, but also make it intriguing. Avoid using too many exclamation marks or words that might trigger spam filters.

Next, personalize your greeting. Even though it's a mass email, using the recipient's name can make a big difference. For instance, instead of saying "Dear Customer," say "Dear Jane." It makes the email feel more personal and less like a generic blast.

Then, dive into the main content. Be concise and direct. People don't have time to read long paragraphs, so get straight to the point. Use bullet points to highlight key information, making it easy to scan. For example, if you're announcing a new product, list its main features and benefits in bullet points.

Also, include clear and compelling calls to action (CTAs). Whether you want your audience to visit your website, make a purchase, or sign up for an event, your CTA should stand out. Use buttons or bold text to make it clear what you want them to do.

Now, about images—yes, they can make your email more appealing, but don't go overboard. Too many images can slow down loading times and might get your email flagged as spam. Use high-quality images that are relevant to your content, and always include alt text in case the images don't load.

Another crucial aspect is mobile optimization. A significant portion of your audience will open your email on their phones. Make sure your email design is responsive, meaning it looks good and functions well on both desktop and mobile devices. Test your emails on various devices to ensure they're easy to read and navigate.

Lastly, before hitting send, thoroughly test your email. Check all links to make sure they work, proofread your content for any typos, and send test emails to yourself and your team. This final step is crucial to catch any errors that might slip through.

Let's not forget about tracking and analyzing the performance of your emails. After you've sent your mass email, use your CRM tool's analytics to track key metrics like open rates, click-through rates, and conversion rates. This data will help you understand what worked and what didn't, allowing you to improve future campaigns.

For example, if you notice that emails sent at a particular time of day have higher open rates, consider scheduling future emails at that time. Or, if certain subject lines perform better than others, analyze them to understand what made them successful and apply those insights to your next email.

Remember, the goal of mass email marketing is to engage your audience and encourage them to take action. By segmenting your audience, crafting compelling content, optimizing for mobile, and continually analyzing your performance, you can create effective mass email campaigns that drive results.

While mass emails are a powerful tool in your marketing arsenal, their success hinges on thoughtful segmentation, clear and engaging content, and meticulous attention to detail. With these strategies, you can ensure your mass emails not only reach your audience but also resonate with them and prompt them to take action.

## Implementing Trigger Emails

The other type of email marketing is trigger emails, which are sent based on specific customer actions or changes in their status. For example, when someone receives an invoice, that's a trigger email. If they abandon their cart, you send a reminder email after a couple of hours.

To effectively implement trigger emails, you need to map out user flows on your website. Identify key actions like checkout, registration, upgrades, and downgrades. Then, pinpoint where users might drop off. For instance, they might add items to their cart but not complete the purchase, or enter their shipping address but leave the page.

Create snapshots of these moments and follow up with users to remind them where they left off. Include a link to bring them back to that point. Often, you'll need to send follow-up emails and occasionally offer discounts. For example, if someone registers but doesn't subscribe immediately, follow up two hours later, one day later, and three days later. As part of this sequence, include promotional coupons to incentivize action.

When creating trigger emails, keep the content relevant and timely. These emails should feel like a natural extension of the user's journey. For instance, if a customer abandons their cart, your email should remind them of the items they left behind and include a clear call to action to complete the purchase. You might say something like, "Hey, you left something in your cart! Click here to finish your purchase."

Personalization is key. Address the recipient by their name and tailor the content to their specific actions. This makes the email feel more personal and less like an automated message. For example, "Hi Sarah, we noticed you were looking at our new collection. Here's a special discount to help you complete your order."

Timing is crucial for trigger emails. You want to follow up soon enough to be relevant but not so quickly that it feels intrusive. For abandoned cart emails, a good strategy might be to send the first reminder a few hours after the cart is abandoned, followed by another reminder the next day. If they still haven't completed the purchase, a final reminder with a discount could be sent a few days later.

Setting up trigger email campaigns involves a few key steps. First, define the triggers—specific actions or inactions that will initiate an email.

Common triggers include:

- Cart abandonment.
- Browsing a particular category or product.
- Registration or sign-up.
- Purchase confirmation.
- Account inactivity.

Once you've defined the triggers, create the email templates. Each template should be concise, clear, and compelling. Use engaging subject lines to capture attention, and ensure the body of the email provides value and a clear call to action.

Next, set up the automation within your CRM tool. Most CRM platforms offer automation features where you can create workflows based on user behavior. Configure these workflows to send the appropriate emails at the right times. For instance, if a user abandons their cart, the workflow should automatically send the first reminder email after a set period.

After setting up your trigger email campaigns, it's essential to monitor their performance. Track key metrics such as open rates, click-through rates, and conversion rates. These metrics will help you understand how well your emails are performing and where you might need to make adjustments.

Regularly review and optimize your emails. Test different subject lines, email content, and send times to see what works best for your audience. A/B testing can be particularly useful here. For instance, you could test two different subject lines to see which one gets a higher open rate, or experiment with different discount offers to see which one drives more conversions.

Implementing trigger emails effectively can significantly enhance your email marketing efforts. By sending timely, relevant emails based on user actions, you can improve engagement and drive conversions. Remember to map out user flows, personalize your content, and continually analyze and optimize your campaigns.

CHAPTER 3

# ALTERNATIVE CHANNELS

Growth hacking is a buzzword you'll hear a lot in marketing circles. It originated in Silicon Valley, and while there are many stories about how it came to be, the concept is pretty straightforward. Growth hacking is all about attracting a large number of new customers to your service or product without significant investment. It's about thinking outside the box and finding innovative ways to grow.

One of my favorite examples of growth hacking is Airbnb. When Airbnb started, they desperately needed more listings. They began by scraping a website called Craigslist, which allowed them to quickly add many listings to their platform. But they didn't stop there. They also worked directly with property owners in key areas like New York, offering to take professional photos of their properties. This not only helped solve the chicken-and-egg problem of needing both listings and users but also made their site visually appealing.

Another great example is Dropbox. When Dropbox launched, users had limited storage space, but they could earn more by inviting friends. This referral program was hugely successful. People started inviting all their friends, and the service quickly gained popularity. In the subscription space, Rent the Runway is another success story. They attribute at least half of their new customers to their referral program.

Growth hacking isn't just about referrals, though. It's about leveraging any available resource to create rapid growth. Take Hotmail, for instance. They added a simple tagline at the bottom of every email sent from their service: "Get your free email at Hotmail." This tiny addition led to massive growth in their user base, showing that sometimes, the simplest ideas can have the biggest impact.

Now, you might be wondering how you can apply growth hacking principles to your business. First, identify your goals. Do you want more users, higher engagement, or increased sales? Once you know what you're aiming for, brainstorm creative ways to achieve those goals without spending a lot of money. Think about your existing resources and how you can use them in new and innovative ways.

One strategy is to create viral content. This doesn't necessarily mean making a funny video that gets millions of views, though that can work. Instead, focus on content that provides real value to your audience and encourages them to share it. For example, a well-researched infographic or a free tool related to your industry can attract attention and drive traffic to your site.

Another approach is to partner with other businesses. Look for companies that complement your own and see if you can collaborate on a project or promotion. This can help you reach a wider audience and gain new customers without significant investment.

You should also consider using data to inform your growth hacking efforts. Analyzing user behavior can reveal insights into what's working and what's not. Use this information to tweak your strategies and maximize your results. For instance, if you notice that a particular type of content gets more engagement, create more of it.

Experimentation is key in growth hacking. Don't be afraid to try new things and see what sticks. Not every idea will work, but the ones that do can lead to significant growth. For example, Twitter experimented with different ways to onboard new users. They found that suggesting accounts for new users to follow right after signing up greatly increased engagement and retention.

Additionally, consider how you can make your product itself a growth driver. Features that encourage users to share or invite others can be very effective. For example, when PayPal first launched, they offered cash incentives for users who referred their friends. This quickly grew their user base and established them as a major player in online payments.

Lastly, always be on the lookout for new trends and opportunities. The digital landscape is constantly changing, and staying ahead of the curve can give you a significant advantage. Whether it's a new social media platform, a technological innovation, or a shift in consumer behavior, being adaptable and open to new ideas is crucial for growth hacking success.

Growth hacking is about creativity, resourcefulness, and a willingness to experiment. By thinking outside the box and leveraging the resources you have, you can achieve rapid growth without breaking the bank. Keep your goals in mind, analyze your data, and stay open to new opportunities. With these strategies, you can take your business to the next level.

## The Power of Influencer Marketing

Influencer marketing has become one of the most visible marketing channels today. It's pretty straightforward: you team up with an influencer in a specific field—like a beauty guru or a popular Twitch streamer—and work together to promote your product. This collaboration can take many forms, such as dedicated videos, product mentions, or unboxing videos.

Unboxing videos, where influencers open and talk about the contents of subscription boxes, are incredibly popular on YouTube. While this format might not work for every product, it's fantastic for companies like Battlebox, which offers a variety of interesting gear. Sometimes, the most unexpected partnerships end up working wonders. For instance, Scentbird had great success by collaborating with comedy influencers who cleverly included mentions of our perfumes in their content.

The ultimate form of influencer collaboration is co-creating a product. A standout example is PewDiePie partnering with YouTube to launch his own TV show. When you're choosing influencers to work with, consider the size of their following. Starting with influencers who have between 5,000 and 10,000 subscribers is cost-effective and often done in exchange for a video. As you grow, targeting influencers with 100,000 to 200,000 subscribers hits a sweet spot for reach and engagement.

Now, let me dive deeper into why influencer marketing is so effective. First off, it's all about trust. Influencers have built a strong rapport with their audience. Their followers trust their opinions and recommendations, which means that when an influencer talks about your product, it feels like a friend is recommending it. This trust translates into higher conversion rates compared to traditional advertising.

Another reason influencer marketing works is the sheer reach it provides. Traditional ads can be easily ignored or blocked, but influencer content is different. Followers actively seek out and engage with it. This means your product is getting in front of people who are genuinely interested in what the influencer has to say. Plus, influencers often have a niche audience that perfectly aligns with your target market, making your marketing efforts even more effective.

Engagement is another critical factor. When influencers promote your product, it's not just a one-time thing. There's usually a conversation that follows. Followers comment, ask questions, and share their thoughts, creating a buzz around your product. This engagement can lead to more visibility and, ultimately, more sales.

One of my favorite examples of successful influencer marketing is how fitness brands collaborate with fitness influencers. These influencers share their workout routines, diet plans, and favorite products with their followers. When they promote a protein powder or a new workout gear, it feels like a genuine recommendation. Their followers see the influencer using the product in their daily lives, which makes it more relatable and trustworthy.

Let's talk about the different types of influencer collaborations in a bit more detail. Dedicated videos are one of the most common forms. In these videos, influencers focus entirely on your product. They might do a review, a tutorial, or a demo. This type of content is great because it provides in-depth information about your product and highlights its features and benefits.

Product mentions are another popular form of collaboration. Here, influencers mention your product in their regular content. It could be a quick shout-out during a vlog or a casual mention in a social media post. This type of collaboration is more subtle and feels less like an ad, which can be more appealing to followers.

Then there are unboxing videos, which I mentioned earlier. These videos are particularly effective for products that come in subscription boxes or have interesting packaging. Influencers film themselves opening the box and exploring its contents, providing their first impressions. This type of content is highly engaging because it taps into the excitement of discovering something new.

Sometimes, brands and influencers take their collaboration to the next level by co-creating products. This is the ultimate form of influencer marketing because it involves the influencer in the product development process. The result is a product that feels authentic and aligned with the influencer's brand. PewDiePie's partnership with YouTube to launch his own TV show is a prime example of this. It wasn't just a promotional campaign; it was a collaboration that created something entirely new and unique.

Choosing the right influencers to work with is crucial for the success of your campaign. As I mentioned earlier, it's often best to start with micro-influencers—those with 5,000 to 10,000 subscribers. These influencers are more accessible and usually willing to collaborate in exchange for free products or a small fee. Despite their smaller following, they often have a highly engaged audience.

As your brand grows, you can start targeting mid-tier influencers with 100,000 to 200,000 subscribers. These influencers offer a balance between reach and engagement. They have a larger following, which means more exposure for your product, but their audience is still engaged and trusts their recommendations.

Of course, working with top-tier influencers—those with millions of followers—can provide massive exposure. However, these collaborations can be expensive and may not always deliver the level of engagement you're looking for. It's essential to weigh the costs and benefits before deciding to work with high-profile influencers.

Influencer marketing is a powerful tool for reaching new audiences, building trust, and driving sales. By partnering with influencers, you can leverage their reach and influence to promote your product in a way that feels authentic and relatable. Whether you're just starting or looking to take your brand to the next level, influencer marketing offers a range of opportunities to connect with your target audience and achieve your marketing goals.

## Exploring Affiliate Marketing

Affiliate marketing is another effective way to attract customers. It's similar to influencer marketing but a bit different. With affiliate marketing, you work with people who create content and include links to your website or product in exchange for a commission. Instead of reaching out to influencers individually, you can collaborate with affiliate platforms that have thousands of registered bloggers and content creators.

Typically, these content creators have smaller followings than major influencers, but their collective reach can be significant. For instance, affiliate marketing can be integrated into computer or phone games, offering incentives like virtual coins for clicking on links. This strategy can drive a lot of traffic to your site. However, it's crucial to monitor for fraud, as some people may try to game the system. Using anti-fraud solutions helps protect your platform and ensures the integrity of your affiliate marketing efforts.

Now, let's dive a bit deeper into the nuts and bolts of affiliate marketing. The basic idea is simple: you find content creators who are willing to promote your product. In return, they get a commission for every sale made through their links. This win-win situation benefits both parties. The content creators earn money, and you get more sales.

One of the main advantages of affiliate marketing is cost-effectiveness. You only pay for actual results. If a content creator generates a sale, they get a commission. If not, you don't lose any money. This performance-based model makes affiliate marketing a low-risk investment compared to other forms of advertising.

Affiliate platforms make the process even more manageable. These platforms connect you with thousands of potential partners, saving you the time and effort of finding and reaching out to each one individually. They also provide tools to track clicks, sales, and commissions, making it easier to manage your affiliate program.

While major influencers have massive followings, the content creators on affiliate platforms usually have smaller audiences. But don't let that fool you. The combined reach of many small content creators can be just as impactful, if not more so, than a single major influencer. Plus, these smaller creators often have highly engaged followers who trust their recommendations.

A great example of this is how some companies integrate affiliate marketing into games. Imagine you're playing a mobile game, and you get an offer to earn virtual coins by clicking a link or watching a short video about a product. This kind of incentive can drive a lot of traffic to your site because players are motivated to engage with the ad.

However, there are challenges to keep in mind. One of the biggest issues in affiliate marketing is fraud. Some individuals might try to game the system to earn commissions without generating legitimate sales. This can include tactics like fake clicks or false leads. To protect your business, it's essential to use anti-fraud solutions. These tools help detect and prevent fraudulent activities, ensuring that your affiliate marketing efforts remain genuine and effective.

Another key aspect of successful affiliate marketing is choosing the right partners. Not every content creator will be a good fit for your brand. It's important to work with affiliates who align with your brand values and target audience. This alignment ensures that the content they create resonates with their followers and drives meaningful traffic to your site.

Communication is also crucial. Maintaining open lines of communication with your affiliates helps build strong relationships and encourages them to promote your products more effectively. Regularly updating them about new products, promotions, and company news keeps them engaged and motivated.

In addition, providing affiliates with high-quality marketing materials can make a big difference. Share banners, product images, and pre-written content that they can use to promote your products. The easier you make it for them to market your products, the more likely they are to do it well.

Tracking and analyzing performance is another essential part of affiliate marketing. Use the tools provided by affiliate platforms to monitor which affiliates are driving the most traffic and sales. This data helps you identify your top performers and optimize your strategy. You can also use this information to reward your best affiliates with higher commissions or special incentives, motivating them to continue promoting your products.

So, there you have it. Affiliate marketing is a powerful tool that can help you attract more customers and drive sales. By partnering with content creators who promote your products in exchange for a commission, you can reach a wider audience and only pay for actual results. Just remember to choose your partners carefully, protect against fraud, and provide your affiliates with the support they need to succeed.

Affiliate marketing offers a range of opportunities to grow your business. Whether you're just starting or looking to expand your reach, this strategy can help you connect with new customers and achieve your marketing goals.

## Utilizing Giveaways

Giveaways have become a go-to strategy in marketing, and for good reason. They offer a fun and engaging way to attract new customers while rewarding your existing ones. The process usually involves partnering with brands that have a similar size and audience to provide a prize, which can be anything from a physical item to a digital gift card.

Participants enter the giveaway by providing their email addresses and, often, sharing the giveaway on social media for extra points. This tactic not only collects valuable email contacts but also increases your brand's visibility across social platforms. It's like hitting two birds with one stone: growing your email list while boosting your online presence.

One of the key benefits of giveaways is their viral nature. They cleverly combine referral marketing and discount programs, creating a buzz that encourages participants to spread the word. As more people share the giveaway, your campaign gains momentum, reaching a much larger audience than you could through traditional advertising methods.

The primary goal of a giveaway is to acquire a large list of new email addresses. But it's crucial to handle these new contacts properly.

Once the giveaway ends, the first step is to clean the email list to ensure all addresses are valid. This might sound tedious, but it's essential to avoid issues with your email deliverability. Invalid or fake email addresses can lead to high bounce rates, which can hurt your sender reputation and even get you flagged as a spammer.

After cleaning the list, you should gradually add these new contacts to your email marketing campaigns. Adding them slowly helps maintain your email deliverability rates and avoids triggering spam filters. You don't want to overwhelm your new subscribers with a flood of emails right away. Instead, introduce them to your brand gradually, starting with a welcome series that explains who you are and what they can expect from your emails.

Now, let's get into some best practices for running a successful giveaway. First, choose a prize that resonates with your target audience. The prize should be something valuable enough to encourage participation but also relevant to your brand. For example, if you run a fitness brand, a giveaway prize could be a bundle of workout gear or a free subscription to a fitness app.

Next, make the entry process simple and straightforward. The easier it is to enter, the more participants you'll get. Asking for an email address is standard, but you can also offer additional entries for actions like sharing the giveaway on social media, tagging friends, or following your brand's social media accounts. These extra actions help increase the giveaway's reach and engagement.

Promoting your giveaway is crucial to its success. Use all available channels to spread the word, including your website, social media, and email newsletters. Collaborating with your giveaway partners on promotion can also extend your reach even further.

Throughout the giveaway, keep the excitement alive by posting updates and reminding your audience to enter. Countdown posts as the end date approaches can create a sense of urgency, encouraging last-minute entries.

Once the giveaway ends, announce the winner publicly. This not only builds trust with your audience but also creates an opportunity to thank everyone who participated and encourage them to stay connected with your brand. Sending a follow-up email to all participants with a special offer or discount can also help convert them into customers.

Giveaways are a powerful tool in your marketing arsenal. They're effective for growing your email list, increasing brand awareness, and engaging with your audience. By choosing the right prize, simplifying the entry process, promoting effectively, and managing new contacts properly, you can maximize the benefits of your giveaway campaigns.

CHAPTER 4

# WEBSITE FUNDAMENTALS

If you're thinking about launching an e-commerce subscription business, you've probably heard of Shopify. It's the go-to platform for anyone wanting to dive into e-commerce. But what's truly fascinating is how Shopify started.

Before 2006, Tobias Lütke, Shopify's current CEO, wasn't in the software business. He actually started a company that sold snowboarding gear online because he loved snowboarding. Back then, they had no tools to run an e-commerce site, so they had to create everything from scratch—website, online payment platform, checkout pages, product pages, and more. They soon realized that building these tools was a bigger opportunity than selling snowboards. Fast forward 18 years, and Shopify has become a pivotal player in the market.

Shopify does more than just help people create online stores. They offer fulfillment services, international launch support, and remote subscription offerings. Plus, there's a huge ecosystem of products built on top of Shopify. This ecosystem lets you fight fraud, accept various payment types, run promotional programs, and sell gift cards. Shopify's creation marked a significant milestone in online commerce.

Now, let's dive a bit deeper into why Shopify is so successful. One of the main reasons is its user-friendly interface. Shopify makes it easy for anyone to set up an online store, even if they have no technical background. The platform provides customizable templates that allow users to create a professional-looking store in no time. This simplicity is a huge draw for small business owners and entrepreneurs who want to get their products online quickly and efficiently.

Another key factor is Shopify's scalability. Whether you're just starting with a small store or running a large enterprise, Shopify can handle it. The platform offers various pricing plans and features that grow with your business. As your store expands, you can upgrade your plan to access more advanced features and support. This flexibility ensures that Shopify remains a valuable tool for businesses at any stage of their growth.

Shopify also excels in providing comprehensive support and resources. Their customer support team is available 24/7 to help users with any issues they might encounter. Additionally, Shopify offers a wealth of educational materials, including blogs, webinars, and tutorials, to help users make the most of the platform. This commitment to customer success is a big part of why Shopify has such a loyal user base.

The integration capabilities of Shopify are another significant advantage. The platform supports numerous third-party apps and services, allowing users to customize their stores to meet their specific needs. Whether you need to integrate with accounting software, email marketing tools, or social media platforms, Shopify has you covered. This extensive integration network makes it easier for businesses to streamline their operations and improve efficiency.

Shopify's marketing tools are also worth mentioning. The platform provides built-in SEO features, social media integration, and analytics tools to help users promote their stores and track performance. These tools enable business owners to make data-driven decisions and optimize their marketing strategies. By providing these resources, Shopify empowers users to grow their online presence and reach a broader audience.

In addition to all these features, Shopify continuously innovates and adapts to changing market trends. The company regularly updates its platform with new features and improvements, ensuring that users always have access to the latest tools and technologies. This commitment to innovation keeps Shopify ahead of the competition and helps users stay relevant in the ever-evolving e-commerce landscape.

Shopify's success story is a testament to its powerful and versatile platform. By providing an easy-to-use interface, scalability, comprehensive support, extensive integration capabilities, effective marketing tools, and top-notch security, Shopify has become the leading choice for e-commerce providers. Whether you're just starting or looking to scale your business, Shopify offers the tools and resources you need to succeed in the online marketplace.

## Overview of E-Commerce Platforms

Shopify isn't the only game in town when it comes to e-commerce platforms. There are numerous other providers out there, each catering to different needs and preferences. Two notable open-source platforms are WooCommerce and Magento. With open-source platforms, you can download the service's original code base, run it on your server, and customize it to fit your needs. This gives you a lot of flexibility, but it also means you're responsible for all the maintenance.

WooCommerce is a plugin for WordPress, making it a great choice if you already have a WordPress site. It's user-friendly and integrates seamlessly with WordPress, providing a smooth transition into e-commerce. WooCommerce is highly customizable, with a vast library of themes and plugins that allow you to tailor your store to your specific needs. However, managing your own server and ensuring security can be challenging if you're not tech-savvy.

Magento, on the other hand, is a more robust and powerful platform. It's designed for larger businesses that need a lot of customization and advanced features. Magento offers a high degree of flexibility, but with that comes complexity. You'll need a strong technical team to manage and maintain your site. For businesses that have the resources, Magento can provide a highly customized and powerful e-commerce solution.

For those who prefer a managed solution, there are platforms like Squarespace, Wix, and WordPress Online. These cloud solutions handle all the maintenance, upgrades, security, and other technical aspects for you, allowing you to focus on running your business. However, they do come with some limitations on customization.

Squarespace is known for its beautiful, responsive templates and ease of use. It's a great option for small businesses or individuals who want a professional-looking site without a lot of hassle. Squarespace provides built-in tools for e-commerce, including inventory management, payment processing, and analytics. The downside is that you're limited to the templates and features provided by Squarespace, which might not be enough for more complex needs.

Wix offers a drag-and-drop website builder that's incredibly user-friendly. It's perfect for those who want to create a website quickly and easily. Wix also offers a variety of e-commerce features, including product galleries, payment processing, and shipping options. However, like Squarespace, Wix has limitations when it comes to customization and scalability.

WordPress Online is another excellent option for those who want a managed solution. It's different from the self-hosted WordPress you use with WooCommerce. With WordPress Online, you get the ease of a managed platform while still having access to a wide range of themes and plugins. It's a good middle ground for those who want flexibility without the hassle of managing their own server.

If you're not a technical founder or don't have a technical team, starting with a cloud solution like Squarespace, Wix, or WordPress Online is advisable. These platforms allow you to test the waters, see what works, and make adjustments as needed without getting bogged down in technical details. As your business grows, you can always consider moving to a more customizable platform like WooCommerce or Magento.

One of the key benefits of these managed solutions is that they take care of security for you. This is a big deal because ensuring your site is secure can be a complex and ongoing task. With a managed solution, you can rest easy knowing that the platform is handling security updates and patches.

Another advantage is the ease of use. Managed platforms are designed to be user-friendly, with intuitive interfaces that make it easy to build and manage your site. You don't need to be a tech expert to create a professional-looking e-commerce site. This is particularly important for small business owners who need to focus on running their business rather than managing a website.

However, it's important to be aware of the limitations. Managed platforms often have restrictions on customization. If your business grows quickly, you might find that you outgrow the capabilities of these platforms. In that case, you might need to transition to a more robust solution like WooCommerce or Magento.

Choosing the right e-commerce platform depends on your specific needs and resources. If you have a technical team and need a highly customizable solution, open-source platforms like WooCommerce and Magento might be the best fit. If you prefer a managed solution that takes care of the technical details for you, platforms like Squarespace, Wix, and WordPress Online offer a great starting point. Evaluate your business needs, consider your technical capabilities, and choose the platform that best supports your goals.

## Crafting Effective User Flows

When you're building a platform—whether it's one I've talked about or something custom—you've got to create a user experience that's both familiar and intuitive. Let's start with the homepage. This is where users land first, and it's vital to make a great impression. You should have multiple versions of your homepage or landing pages tailored to different advertising sources. These pages are the gateway to your product, setting the stage for everything that follows.

Next, product discovery is super important. Visitors need to find what they're looking for easily, whether it's a range of products or just one specific item. Your product pages, or PDPs, should clearly describe the product, what's included, and why customers will love it. Don't forget to include subscription selection pages and account pages where users can manage their subscriptions.

The checkout process is another crucial area. It should be seamless and straightforward. Whether it's for subscriptions, e-commerce purchases, or gifts, each flow should be smooth and hassle-free. The account profile page should be user-friendly, allowing users to manage their personal information with ease.

And let's not overlook post-purchase pages like tracking pages. These are essential for a complete user experience, ensuring customers can follow their orders and feel confident in their purchase. Alright, let's dive into some more specifics.

## Homepage and Landing Pages

Your homepage isn't just the front door to your website; it's your first opportunity to wow your visitors. Think of it like the entrance to a store—it should be welcoming, engaging, and make people want to explore more. You should have different versions of your homepage, also known as landing pages, to cater to various advertising sources. For instance, a user who clicks through from a social media ad might be looking for something different than someone who arrives via a search engine. Tailoring these pages helps ensure that each visitor gets the best first impression possible.

When visitors arrive on your site, they should immediately know how to find what they're looking for. This means having a clear, intuitive navigation system and a search function that actually works. If you're selling multiple products, make sure each category is easy to locate. If you're offering a single product, your product page should shine. This page should provide a thorough explanation of the product, highlighting what's inside and why customers will love it. High-quality images, detailed descriptions, and customer reviews can go a long way in convincing potential buyers.

### Subscription Selection and Account Management

Subscription services are becoming increasingly popular, and for good reason—they provide a steady revenue stream and keep customers coming back. Your subscription selection page should be straightforward, allowing users to easily choose the plan that best fits their needs. Once they've subscribed, the account page should offer an intuitive interface for managing their subscription. Users should be able to update their payment information, change their subscription plan, and view their billing history without any hassle.

### Checkout Process

A smooth checkout process is vital for converting visitors into paying customers. Any hiccups here can result in abandoned carts and lost sales. Your checkout flow should be as simple as possible, with clear instructions and minimal steps. Offer different checkout processes for different types of purchases—subscriptions, single products, and gifts. This tailored approach can help reduce confusion and make the process more efficient.

### Account Profile Page

Once customers have made a purchase, they should have easy access to their account profile. This page should allow them to manage their personal information, view their order history, and track their current orders. It's important that this page is user-friendly and secure, giving customers confidence that their data is safe.

## Post-Purchase Pages

The experience doesn't end once the purchase is made. Post-purchase pages like order confirmation and tracking pages are essential for providing a complete user experience. These pages reassure customers that their order is being processed and give them the ability to track their shipment. Providing clear and timely updates can help build trust and encourage repeat business.

Crafting effective user flows is about making the entire journey—from landing on your homepage to post-purchase tracking—as smooth and enjoyable as possible. By focusing on these key areas, you can create a user experience that not only meets but exceeds customer expectations. Remember, a happy customer is a returning customer, and these user flows are your roadmap to achieving that.

## Utilizing Product Analytics

Once your website is live, it's not just a static entity. You'll need to keep experimenting to see what works and what doesn't. This is where product analytics come into play. Analytics tools help you understand user behavior by tracking the paths users take on your site, often referred to as funnels. For instance, a simple funnel might look like this: homepage > registration page > checkout page > thank you page. These tools show you how many people complete each step, highlighting areas where you can improve.

One of the most powerful aspects of product analytics is the ability to see exactly how users interact with your site. Many analytics services record user sessions, offering insights into customer behavior without compromising privacy. These recordings let you see how users navigate your site, fill out forms, and interact with different elements. You might be surprised to find that actual user behavior often differs from what you expected.

Heatmap services are another valuable tool in your analytics arsenal. Heatmaps show you how far users scroll on your pages, where they click, and which parts of the page get the most attention. This data is incredibly useful for understanding what engages your audience and where you might need to make changes.

Alright, let's dig deeper into each of these tools and how they can help you improve your website.

### Understanding Funnels

Funnels are a key concept in product analytics. They represent the path that users take on your website, from the moment they land on your homepage to the final step of completing a purchase or signing up for a service. By analyzing these funnels, you can see where users drop off and where they continue to the next step. This information is crucial for identifying bottlenecks and optimizing your site's performance.

For example, if you notice a large drop-off between the registration page and the checkout page, it might indicate that your registration process is too complicated or that there's something on the registration page that's turning users away. By pinpointing these problem areas, you can make targeted changes to improve the user experience and increase conversions.

### Recording User Sessions

Recording user sessions is like having a window into the actual user experience. These recordings let you see exactly how users interact with your site in real-time. You can watch as users navigate through your pages, fill out forms, and click on different elements. This firsthand view can reveal a lot about user behavior and preferences.

Often, you'll find that users don't behave the way you expected. Maybe they're getting stuck on a particular form field, or perhaps they're missing a crucial call to action because it's not visible enough. By watching these recordings, you can identify these issues and make the necessary adjustments to improve the user experience.

### Utilizing Heatmaps

Heatmaps are another fantastic tool for understanding user behavior. They provide a visual representation of where users click, how far they scroll, and which parts of the page get the most attention. This information can help you understand what's working well on your site and what needs improvement.

For instance, if you notice that users are clicking on a non-clickable element, it might indicate that they're expecting it to be interactive. Similarly, if a crucial call to action isn't getting any clicks, it might be because it's not prominently placed or visually appealing. Heatmaps can also show you how far users are scrolling down your pages. If users are not scrolling far enough to see important content, you might need to move that content higher up on the page.

### Making Data-Driven Decisions

The ultimate goal of using product analytics is to make data-driven decisions that improve your website's performance and user experience. By understanding how users interact with your site, you can make informed changes that lead to better engagement and higher conversions.

For example, let's say you're running an e-commerce site and you notice that a significant number of users are abandoning their carts on the checkout page. By analyzing your funnels, session recordings, and heatmaps, you might discover that the checkout process is too lengthy or that there are technical issues causing frustration. Armed with this information, you can streamline the checkout process, fix any technical issues, and ultimately reduce cart abandonment rates.

## Continuous Improvement

Product analytics isn't a one-time task. It's an ongoing process of monitoring, analyzing, and optimizing your website. User behavior can change over time, and what works today might not work tomorrow. By continuously tracking and analyzing user interactions, you can stay ahead of the curve and keep improving your site.

Utilizing product analytics is essential for creating a successful and user-friendly website. By understanding how users navigate your site, where they click, and how they interact with different elements, you can make informed decisions that enhance the user experience and drive better results. So, keep experimenting, stay curious, and let the data guide your way to success.

## The Importance of A/B Testing

Once you have your analytics, you'll want to make improvements. But how can you be sure your changes will work? That's where A/B testing comes in. A/B testing lets you compare two versions of a page to see which performs better. You split your traffic so half sees version A and half sees version B.

You can test various elements, from copy and images to layout and user flows. For example, you might find that users prefer "auto-ship" over "subscribe" because it feels less like a commitment. You can test visuals, like different ways to present a product, which can significantly impact engagement. Even small changes, like font size and button placement, can make a difference.

In some cases, you might test user flows. For example, should users register before checkout? Guest checkouts are popular, but if you don't capture an email, you can't follow up with abandoned carts. A/B testing helps you determine which approach works best for your goals.

A/B testing, also known as split testing, is like a scientific experiment for your website. It involves creating two versions of a webpage—the original (A) and a modified version (B). By randomly showing these versions to different segments of your visitors, you can see which one performs better in terms of your desired outcome, such as click-through rates, conversions, or time spent on the page.

A/B testing is crucial because it takes the guesswork out of website optimization. Instead of making changes based on hunches or assumptions, you can rely on real data to guide your decisions. This method helps you understand what your audience responds to best, ensuring that your updates lead to actual improvements.

For example, let's say you want to increase the number of users who complete a purchase on your site. You might hypothesize that changing the call-to-action (CTA) button from "Buy Now" to "Shop Now" will make a difference. Instead of just making the change and hoping for the best, you can use A/B testing to compare the two CTAs and see which one drives more conversions.

You can A/B test almost any element on your website, but some common areas include:

Headlines and Copy: The text on your website, especially headlines, can significantly impact user engagement. Test different wording, lengths, and tones to see what resonates most with your audience.

Images and Videos: Visual content is powerful. Try different images or videos to see which ones capture attention and encourage action.

Layout and Design: The overall design and layout of your pages can affect how users interact with your site. Test different arrangements of elements, like the placement of the CTA button or the navigation menu.

Call-to-Action Buttons: The text, color, size, and placement of your CTA buttons can influence whether users click on them. Experiment with different variations to find the most effective combination.

Forms: If your site includes forms, such as registration or contact forms, test different lengths, fields, and layouts to improve completion rates.

## Running an A/B Test

Running an A/B test involves several steps:

Identify a Goal: Decide what you want to improve, such as increasing sign-ups, reducing bounce rates, or boosting sales.

Create Variations: Develop the different versions of the element you want to test. For example, if you're testing a headline, create two or more versions.

Split Your Traffic: Use an A/B testing tool to randomly divide your traffic between the original version (A) and the new version (B).

Collect Data: Allow the test to run for a sufficient amount of time to gather enough data. The duration will depend on your traffic volume and the nature of the change.

Analyze Results: Compare the performance of the two versions based on your goal. Look at metrics like conversion rates, click-through rates, and engagement levels.

Implement the Winner: Once you have a clear winner, implement the successful version on your site.

## Examples of A/B Testing in Action

Here are a few real-world examples of how A/B testing can lead to significant improvements:

Button Text: A company tested two versions of a CTA button—one that said "Get Started" and another that said "Sign Up Now." The "Get Started" button led to a 20% increase in conversions because it felt less committal.

Image Placement: An online retailer tested the placement of product images on their homepage. By moving the images higher on the page, they saw a 15% increase in clicks on product links.

Form Length: A software company tested two versions of a sign-up form—one long and one short. The shorter form resulted in a 30% increase in submissions, as users were more likely to complete it.

A/B testing is an invaluable tool for website optimization. By systematically testing different elements of your site, you can make data-driven decisions that lead to better performance and higher user satisfaction. So, start experimenting, track your results, and watch your website's effectiveness grow.

CHAPTER 5

# PLANS AND PRICING

When you're ready to launch your business, you've got the website set up, and everything is in place. Now comes a crucial but often tricky part: pricing your subscription. This is where human psychology comes into play. One of the greatest books that describe this is "Nudge," which explains how our brains make choices.

The book shares many examples, like rearranging food in school cafeterias so kids choose healthier options over pizza or pasta. There are also experiments on how to organize pricing on your website. There's a reason we usually see three options when we go to any SaaS service or subscription box site.

A famous experiment conducted by the New York Times illustrates this well. Initially, they had two pricing options—one for $10 and one for $20. When they introduced a middle option for $15, they saw an increase in conversions. Surprisingly, more people started choosing the most expensive option. This is because the $15 option made the $20 choice seem like a better deal by comparison.

When setting up your subscription pricing, it's ideal to offer multiple payment or subscription options. This doesn't just mean monthly plans; consider three, six, or twelve-month plans as well. The exact number of options depends on what works best for your business, but three is usually a good starting point.

The more expensive plans should offer added value. For instance, a twelve-month plan could offer one month free, making it effectively twelve months for the price of eleven. This not only makes the higher-tier plans more attractive but also encourages customers to commit for a longer period, increasing your customer retention and reducing churn.

When calculating the price, start with the unit economics of your subscription. Understand the cost of goods and customer acquisition costs. Aim to become profitable by month two or three, and target at least a 50% gross margin. This approach leads you to the right subscription price.

Here's a step-by-step approach to help you determine your pricing:

Calculate Cost of Goods Sold (COGS): This includes the cost of the products or services you're offering, packaging, and shipping.

Determine Customer Acquisition Cost (CAC): This is the cost associated with acquiring a new customer, including marketing and advertising expenses.

Set a Profit Margin: Aim for at least a 50% gross margin. This means if your COGS and CAC add up to $20, your subscription price should be at least $40.

Factor in Discounts and Promotions: If you're offering discounts for longer-term subscriptions, make sure these are factored into your pricing strategy so they don't erode your profit margins.

The Importance of Testing and Iteration

Pricing is not a set-it-and-forget-it task. You need to continuously test and iterate to find the sweet spot that maximizes both conversions and revenue. Use A/B testing to compare different pricing strategies and see what resonates best with your audience.

For instance, we initially offered a flat monthly subscription. After testing, we introduced three, six, and twelve-month plans with discounts for longer commitments. We found that a significant portion of our customers opted for the six-month plan, boosting our revenue and reducing churn.

## Utilizing Psychological Triggers

Using psychological triggers can also enhance your pricing strategy. Here are a few tactics:

Anchoring: Show the highest-priced plan first to make the other options seem more affordable.

Decoy Pricing: Introduce a middle option that makes the higher-priced plan seem like a better value.

Charm Pricing: Use prices that end in .99, as they tend to be perceived as significantly lower than a round number.

## Real-World Example: Netflix

Netflix is a great example of a company that uses tiered pricing effectively. They offer three plans: Basic, Standard, and Premium. Each plan offers different features, such as the number of screens that can be watched simultaneously and the availability of HD streaming. This tiered approach allows Netflix to cater to different customer needs while maximizing their revenue.

Pricing your subscription service effectively is crucial for your business's success. By understanding the psychology behind pricing, offering value with different plans, and continuously testing and iterating your strategy, you can find the optimal price points for your service. Remember to factor in your costs, aim for a healthy profit margin, and utilize psychological triggers to encourage higher-tier subscriptions. With the right approach, you can create a pricing strategy that not only attracts customers but also drives long-term growth and profitability.

## Designing Your Subscription Plans

Designing subscription plans involves considering cost, price, and value. Sometimes, customers might not see enough value in your subscription, even if they can afford it. Think of luxury items like handbags. For example, Hermès bags are made of high-quality leather by skilled artisans, but their price far exceeds the cost of materials and labor. The perceived value is what drives the high price.

When designing subscription plans, ensure customers see value in what they're getting for the price they pay. Offer various plans, such as monthly, three-month, and six-month options. The duration depends on how customers consume your product. For instance, a twelve-month plan might be suitable for shaving supplies or supplements, where effects are seen over time.

## Balancing Cost, Price, and Value

To design effective subscription plans, you must balance cost, price, and perceived value. Start by understanding your cost of goods sold (COGS), which includes everything from production to shipping. Then, set a price that covers these costs while also providing a healthy profit margin. But more importantly, ensure that the value customers perceive aligns with what they pay.

## Understanding Customer Perception

Perceived value is a critical factor in pricing. Customers must feel they're getting more than just the physical product—they should feel they're receiving something special or exclusive. This could be through high-quality products, exceptional customer service, or unique benefits like exclusive access to content or early product releases.

## Offering Multiple Plans

Offering multiple subscription plans caters to different customer needs and preferences. For instance, you could offer:

Monthly Plans: Ideal for customers who prefer flexibility and may be hesitant to commit long-term.

Three-Month Plans: Often come with a small discount, enticing customers to commit for a slightly longer period.

Six-Month or Annual Plans: These should offer the best value, perhaps with a significant discount or additional perks. For example, a twelve-month subscription could offer one month free, making it effectively eleven months for the price of twelve.

## Matching Plans to Product Use

The subscription duration should match how customers use your product. For example, if you're selling shaving supplies, customers might see the value in a twelve-month plan since the benefits of consistent use (like improved skin health) are seen over time. Similarly, for supplements, a longer subscription makes sense because it often takes a few months to notice significant health benefits.

## Pricing Strategies

Pricing your subscription plans involves more than just covering costs. It's about finding the sweet spot where customers feel they're getting good value for money. Here are some pricing strategies to consider:

Competitive Pricing: Look at what competitors charge for similar products and position your pricing accordingly. You don't always have to be the cheapest, but you should offer comparable or superior value.

Value-Based Pricing: Price your subscriptions based on the value they provide to customers. This might mean higher prices for premium features or exclusive content.

Psychological Pricing: Use pricing tactics like charm pricing (e.g., $9.99 instead of $10) to make prices seem lower. Offering a high-priced premium option can also make your standard plans seem more affordable.

## Enhancing Perceived Value

To increase the perceived value of your subscription, consider offering:

Exclusive Content: Access to exclusive articles, videos, or webinars that are not available to non-subscribers.

Community Access: Create a community around your product where subscribers can interact, share tips, and support each other.

Personalization: Tailor your products or services to meet the specific needs of each subscriber. This could involve personalized recommendations or custom products.

### Real-World Example: Blue Apron

Blue Apron, a meal kit delivery service, offers various subscription plans to cater to different customer needs. They have two-person plans and family plans, with options to receive two, three, or four recipes per week. They also provide flexibility for customers to skip weeks if needed. By offering different plans and customization options, Blue Apron can attract a broader range of customers.

## Testing and Iteration

After launching your subscription plans, continuously test and iterate to optimize pricing and offerings. Use A/B testing to see which pricing strategies and plan structures work best. Collect feedback from your customers to understand what they value most and adjust your plans accordingly.

For instance, if you notice that a large number of customers prefer the three-month plan, you might consider adding more value to the six-month plan to make it more attractive. Regularly reviewing your subscription metrics can help you identify trends and areas for improvement.

Designing your subscription plans is a crucial step in building a successful subscription business. By balancing cost, price, and value, understanding customer perception, and offering various plans that match product use, you can create compelling subscription options. Use competitive, value-based, and psychological pricing strategies to enhance perceived value. Continuously test and iterate your plans to find the optimal pricing and offerings. With the right approach, you can attract and retain customers, ensuring long-term success for your subscription business.

## Pricing Your Subscription Plans

Pricing is a critical aspect because, while you want your service to be accessible, you're also running a business to make a profit. Here are key factors to consider:

### Cost of Goods

Calculate how much it costs to create one subscription box. This includes the cost of products, packaging, and shipping. As your business grows, you'll likely get better deals on items, but focus on profitability from the start. Aim to break even or make a small profit initially. Knowing your costs inside and out helps you set a price that covers expenses and generates profit.

### Customer Acquisition Costs (CAC)

Customer Acquisition Costs (CAC) include all the expenses associated with acquiring new customers, such as marketing, advertising, and promotional campaigns. Initially, your marketing expenses might be high as you test different strategies and channels to see what works best. Over time, you'll optimize these efforts and reduce your CAC. Understanding your CAC is essential for pricing your subscription so that you can cover these costs and still make a profit.

### Margin

In traditional commerce, profit margins are often thin, but digital SaaS (Software as a Service) businesses can have margins as high as 90-97%. For subscription boxes, a target margin of 70-80% is ideal. Starting with a 50% margin provides a solid foundation, allowing room for growth and unexpected costs. Over time, aim to increase this margin by optimizing operations, reducing costs, and offering premium options.

## Understanding Customer Value

When setting prices, consider the value your customers get from your product or service. Ensure they perceive the subscription as worth the cost. Customers need to feel they're getting a good deal and that the benefits of the subscription outweigh the price. Here are some tips to enhance perceived value:

Quality Products: Ensure the items in your subscription box are of high quality. Customers are more likely to see the value in their subscription if they receive products they love and find useful.

Exclusive Items: Include items that are exclusive to your subscription box. This exclusivity can make customers feel they're getting something special that they can't find elsewhere.

Personalization: Personalize the subscription experience. Tailored recommendations and customized boxes based on customer preferences can significantly enhance perceived value.

Excellent Customer Service: Provide top-notch customer service. Quick responses to inquiries and a hassle-free return policy can make customers feel valued and increase their perception of your subscription's worth.

## Real-World Example: FabFitFun

FabFitFun is a great example of a subscription box that understands pricing and customer value. They offer a seasonal box filled with full-sized products across beauty, fitness, wellness, and home.

Their boxes cost around $50 per season, but the total value of the items inside often exceeds $200. This substantial perceived value makes customers feel they're getting a great deal, encouraging them to stay subscribed.

## Adjusting Prices Over Time

As your business grows and you gain a better understanding of your costs and customer preferences, you might need to adjust your prices. Here are some scenarios where you might consider a price adjustment:

Increased Costs: If your cost of goods or shipping increases, you might need to raise your prices to maintain your profit margin.

Enhanced Value: If you've significantly improved the quality of your products or added new features, you can justify a higher price.

Market Changes: Keep an eye on your competitors. If the market changes and your competitors adjust their prices, you might need to follow suit to stay competitive.

Customer Feedback: Listen to your customers. If they consistently tell you that your subscription is too expensive or too cheap, consider adjusting your prices accordingly.

## Testing Pricing Strategies

Pricing isn't a one-size-fits-all approach. It's important to test different pricing strategies to find what works best for your business and your customers. Here are a few strategies to consider:

Bundling: Offer bundled packages at a discounted rate. For example, a three-month subscription for a lower monthly price than a single month.

## Transparency and Communication

When adjusting prices, transparency and communication are key. Be upfront with your customers about why you're changing the price and how it will benefit them. Whether it's due to increased costs, improved products, or additional features, clear communication helps maintain trust and customer loyalty.

Pricing your subscription plans is a critical aspect of running a successful subscription business. By considering your cost of goods, customer acquisition costs, and desired profit margins, you can set a price that covers your expenses and generates profit. Always ensure that customers perceive value in what they're getting for the price they pay. Regularly review and adjust your pricing based on market changes, customer feedback, and your evolving business needs. With thoughtful pricing strategies, you can attract and retain customers while ensuring the long-term success of your business.

CHAPTER 6

# ONLINE PAYMENTS

In today's culture, we're so used to paying for goods and services online. We just take out our credit cards, type in the information, and it works. But this convenience is a relatively recent invention. Back in the 1980s, there was no such thing as a credit card.

The story of Visa, a pivotal player in electronic payments, is fascinating. While they weren't the only ones to come up with an electronic payment solution, they are the biggest. It all started with a meeting of regional participants of Bank of America. They introduced the idea of electronic payments, and it quickly became a success because people could pay for things online. Initially, only credit cards were created, but eventually, banks also introduced debit cards.

Today, we have not just Visa but also MasterCard, American Express, and Discover. Visa's technology is remarkable, handling millions of transactions daily. They pioneered many concepts in software engineering, including loyalty cards and cashbacks. This relatively small organization created the backbone of modern payments worldwide.

In the 1980s, the concept of electronic payments was revolutionary. People were used to cash and checks, and the idea of a plastic card storing your financial information seemed futuristic. However, a group of innovative thinkers saw the potential in creating a system where transactions could be conducted electronically, offering convenience and security.

Visa started as an ambitious project by a group of regional bank participants. Their goal was to streamline payments and make transactions faster and safer for consumers and businesses alike. They began by issuing credit cards, which allowed people to buy now and pay later, a concept that quickly gained popularity.

As the idea of electronic payments took off, Visa didn't stop at credit cards. They expanded their offerings to include debit cards, which allowed customers to pay directly from their bank accounts. This move broadened their user base and made electronic payments accessible to a larger audience.

Visa's commitment to innovation didn't end with expanding their card offerings. They also pioneered several concepts that are now standard in the payment industry. For example, they introduced loyalty programs and cashback offers, which incentivized customers to use their cards more frequently. These programs not only increased card usage but also helped build customer loyalty.

The technology that powers Visa's transactions is nothing short of remarkable. Every day, Visa processes millions of transactions worldwide, ensuring that each one is completed quickly and securely. This requires sophisticated software and robust infrastructure to handle the massive volume of data.

Visa's technology team developed cutting-edge systems to manage these transactions. They created secure networks to protect customer information and prevent fraud. They also implemented advanced algorithms to detect suspicious activity and safeguard users' funds. This focus on security and reliability has made Visa a trusted name in the payment industry.

Visa's influence extends far beyond the realm of electronic payments. By making transactions easier and more efficient, they have played a crucial role in shaping the modern global economy. Businesses of all sizes benefit from the ability to accept electronic payments, which can increase sales and improve cash flow.

For consumers, Visa has made purchasing goods and services more convenient than ever. Whether shopping online or in-store, customers can complete transactions quickly and securely with a Visa card. This ease of use has contributed to the growth of e-commerce and transformed how people shop.

While Visa is a dominant player in the electronic payments industry, they are not alone. MasterCard, American Express, and Discover are also significant players, each offering their own unique features and benefits. This competition drives innovation and ensures that consumers have various options for their payment needs.

MasterCard, for instance, has focused on expanding its global reach and offering unique perks like travel rewards and concierge services. American Express is known for its premium cards and exceptional customer service. Discover has made a name for itself with generous cashback offers and no annual fees. This competitive landscape keeps all players striving to improve their offerings and better serve their customers.

The future of electronic payments is bright, and Visa continues to lead the way with ongoing innovation. Contactless payments, digital wallets, and mobile payment solutions are becoming increasingly popular, and Visa is at the forefront of these developments. They are continually exploring new technologies to make transactions even more seamless and secure.

One exciting area of growth is the use of blockchain technology and cryptocurrencies. While still in its early stages, these innovations have the potential to revolutionize the payment industry further. Visa is actively researching and investing in these areas to stay ahead of the curve and provide the best possible service to their customers.

Understanding Visa's journey helps us appreciate the convenience we often take for granted. As we swipe our cards or tap our phones to pay, we are participating in a sophisticated system that has been decades in the making. The story of Visa is not just about a company but about the evolution of commerce and the ongoing quest to make our financial lives easier and more secure.

When you or a customer hits the "Pay Now" button on a website, it triggers a sophisticated process that varies based on the payment method used. Let's start with credit cards.

First, the credit card information is tokenized and submitted to a payment provider. This provider collects all the data and sends it to Visa or another relevant network. The data then goes to the bank that issued the card. If the customer has enough funds and the transaction isn't flagged as fraudulent, the bank approves the transaction. This approval message travels back through the payment provider to the merchant, indicating that the money is secured. However, the funds don't immediately hit the merchant's bank account; reconciliation happens at the end of the day, with funds transferred the next day.

## Tokenization

Tokenization is a security measure that replaces sensitive credit card information with a unique identifier or "token." This token is then used to process the payment without exposing the actual credit card details, reducing the risk of data breaches and fraud.

### Payment Providers

Payment providers, such as Stripe, Square, or PayPal, play a crucial role in the transaction process. They act as intermediaries between the merchant and the credit card networks, ensuring that the transaction data is securely transmitted and processed. They also handle various compliance and security requirements, making it easier for merchants to accept online payments.

### Apple Pay, Google Wallet, and Amazon Pay

Other payment methods include Apple Pay, Google Wallet, and Amazon Pay. These aren't payment methods themselves but rather convenient ways to pay online. They store the user's payment information securely and use tokenization to process transactions quickly and safely.

### Apple Pay

Apple Pay uses near-field communication (NFC) technology for in-store payments and tokenization for online transactions. When a user makes a purchase, Apple Pay creates a unique transaction code and processes the payment without sharing the actual card details with the merchant.

### Google Wallet

Google Wallet works similarly, allowing users to store their credit and debit card information securely. It uses tokenization and encryption to ensure the safety of online transactions, providing a seamless and secure payment experience.

## Amazon Pay

Amazon Pay lets users pay using their Amazon account information. It streamlines the checkout process by eliminating the need to enter payment and shipping details for each transaction, leveraging the information already stored in the user's Amazon account.

## PayPal

PayPal, introduced in 2000, is a unique payment method that works quickly without intermediate players. It uses its own funds, making transactions faster. When a customer pays with PayPal, the funds are transferred from the customer's PayPal account to the merchant's PayPal account almost instantly. PayPal then settles the transaction with the customer's bank or credit card company.

## Buy Now, Pay Later Services

Buy Now, Pay Later (BNPL) services, like Klarna, Afterpay, and Affirm, have become increasingly popular. These services allow customers to split their purchases into smaller, interest-free payments over time. When a customer uses a BNPL service, the provider pays the merchant in full upfront and then collects the installment payments from the customer.

## Regional Payment Options

In addition to global payment methods, there are regional options like WeChat Pay in Asia. WeChat is not just a payment method but has an underlying business logic that facilitates transactions. WeChat Pay integrates seamlessly with the WeChat app, allowing users to make payments, transfer money, and pay bills directly within the app.

## WeChat Pay

WeChat Pay uses QR codes for in-store payments and tokenization for online transactions. Users link their bank accounts or credit cards to their WeChat accounts and can make payments by scanning QR codes or entering a PIN.

## Potential Issues in the Payment Ecosystem

This ecosystem is quite fragile, with many things that can go wrong. Understanding potential issues helps you ensure smooth transactions for your business.

## Common Issues and Solutions

Declined Transactions: Transactions can be declined for various reasons, such as insufficient funds, expired cards, or suspected fraud. To minimize declines, ensure your payment gateway supports multiple attempts and provides clear error messages to customers.

Fraud Prevention: Fraud is a significant concern in online payments. Implement security measures like tokenization, encryption, and two-factor authentication to protect your customers' data and reduce the risk of fraudulent transactions.

Chargebacks: Chargebacks occur when customers dispute a transaction and request a refund from their bank. To reduce chargebacks, provide clear transaction details, excellent customer service, and a straightforward refund policy.

Integration Issues: Integrating multiple payment methods can be complex. Use reliable payment gateways and APIs to ensure seamless integration and minimize technical issues.

Currency Conversion: For international transactions, currency conversion can be a challenge. Ensure your payment provider supports multiple currencies and offers competitive exchange rates.

Understanding how online payments work is crucial for running a successful e-commerce business. From credit card payments to digital wallets and regional payment options, each method has its unique processes and potential issues. By implementing robust security measures, optimizing your payment gateway, and staying informed about the latest payment technologies, you can ensure smooth and secure transactions for your customers. This not only enhances their experience but also builds trust and loyalty, driving the long-term success of your business.

Subscription Billing Platforms and Payment Providers

Starting a subscription business involves many considerations. You need to accept various payment methods, run transactions regularly, create discounts, recover charges, and ensure security and compliance. Compliance includes adhering to standards like PCI DSS, which requires stringent measures for online payments.

Implementing all these features from scratch is challenging unless you're a large company like Airbnb. Fortunately, there's an entire industry of subscription billing platforms designed to simplify this process. Companies like Recurly, ChargeBee, and Recharge (especially for Shopify users) provide convenient platforms for running subscription businesses.

Subscription billing platforms offer comprehensive solutions to manage various aspects of a subscription business. They allow you to create different subscription plans, manage discounts, store customer profiles, and handle recurring transactions seamlessly.

One of the first steps in setting up a subscription business is creating subscription plans. These platforms let you easily set up and customize different plans to cater to various customer needs. You can offer monthly, quarterly, or annual subscriptions, and even provide options for different tiers of service.

For example, if you're running a subscription box service, you might offer a basic plan with standard items and a premium plan with exclusive products. These platforms make it easy to manage and modify these plans as your business grows.

### Managing Discounts

Offering discounts and promotions is a great way to attract new customers and retain existing ones. Subscription billing platforms allow you to create and manage discounts effortlessly. You can set up discount codes, run special promotions, and track their effectiveness.

We've found that running limited-time promotions, like a 20% discount for the first three months, significantly boosts our subscriber base. The ability to manage these promotions seamlessly through our billing platform saves us a lot of time and effort.

### Handling Recurring Transactions

One of the core functions of subscription billing platforms is managing recurring transactions. These platforms automate the billing process, ensuring that customers are charged correctly and on time. This automation reduces the administrative burden and helps maintain a steady cash flow.

### Storing Customer Profiles

Customer profiles are essential for managing subscriptions effectively. These platforms store detailed information about each customer, including payment details, subscription preferences, and transaction history. This data helps in providing personalized service and addressing any issues promptly.

For instance, if a customer needs to update their payment information or change their subscription plan, the platform makes it easy to access and modify their profile. This streamlined process enhances the customer experience and reduces churn.

### Dunning Management

An essential feature of any subscription billing platform is dunning management, which helps recover failed charges. There are many reasons a charge can fail, such as insufficient funds, fraud issues, or technical problems at any point in the payment chain. The expertise and tools provided by subscription billing platforms are invaluable in managing these challenges.

### Recovering Failed Charges

Dunning management involves retrying failed transactions and communicating with customers to resolve payment issues. Advanced platforms use intelligent algorithms to optimize the timing of retries, increasing the chances of successful payments.

In our experience, about 20% of initial payment attempts fail, but we recover around 80% of those failed charges through effective dunning management. This recovery significantly impacts our bottom line and reduces involuntary churn.

### Security and Compliance

Security and compliance are critical in the subscription business. Subscription billing platforms adhere to strict security standards like PCI DSS, ensuring that customer payment information is handled securely. Compliance with these standards is essential to protect your business from fraud and legal issues.

## PCI DSS Compliance

PCI DSS (Payment Card Industry Data Security Standard) compliance involves implementing stringent security measures to protect cardholder data. This includes encrypting payment information, maintaining secure networks, and regularly monitoring and testing systems.

By using a subscription billing platform that complies with PCI DSS, you ensure that your business meets these security requirements. This not only protects your customers but also builds trust and credibility for your brand.

## Choosing the Right Platform

Choosing the right subscription billing platform is crucial for the success of your subscription business. Here are some factors to consider:

Integration: Ensure the platform integrates seamlessly with your existing systems, such as your e-commerce platform, CRM, and accounting software.

Scalability: Choose a platform that can scale with your business. As your subscriber base grows, the platform should be able to handle increased transaction volumes without compromising performance.

Support and Reliability: Look for a platform with robust customer support and a track record of reliability. You need a partner you can count on to address issues promptly and keep your operations running smoothly.

Cost: Evaluate the pricing structure to ensure it fits within your budget. Consider both the upfront costs and ongoing fees, and weigh them against the features and benefits the platform offers.

Real-World Example: Using Recharge for Shopify

For Shopify users, Recharge is a popular choice. It integrates seamlessly with Shopify, allowing you to manage subscriptions directly from your Shopify dashboard. Recharge offers a range of features, including customizable subscription plans, dunning management, and detailed analytics.

We chose Recharge for one of our Shopify stores because of its ease of use and robust feature set. The integration was straightforward, and the platform's capabilities have helped us manage our subscriptions more effectively. The detailed analytics provide valuable insights into our subscription business, helping us make data-driven decisions.

Subscription billing platforms are essential tools for managing a successful subscription business. They simplify the process of creating subscription plans, handling recurring transactions, managing discounts, and ensuring security and compliance. By choosing the right platform, you can streamline your operations, reduce administrative burdens, and focus on growing your business. The right platform not only makes it easier to manage subscriptions but also enhances the customer experience, ultimately driving the success of your subscription business.

CHAPTER 7

# POST-PURCHASE EXPERIENCE

Every year, hundreds of packages get lost. For a shipping company, it's just another entity they pay an insurance fee for. But for customers, it's a big deal—they didn't get anything. For merchants, it's also a significant loss. Not only do they lose money on the transaction, but they also have to deal with the claim and send a replacement to the customer, which results in additional costs without generating revenue.

Understanding the reasons behind lost packages is crucial when running a business. Packages can get lost due to issues on the carrier's side, like a slip-up or truck damage. Sometimes, it's due to customers moving and not updating their address with USPS or the equivalent service in their country. In such cases, the package gets delivered to the old address and might never reach the intended recipient. It could end up in the hands of someone else or get processed as a return, incurring additional costs.

Managing lost packages is part of ensuring a great customer experience. It's vital to pay close attention to this aspect, as it directly impacts how customers perceive your service.

## The Impact of Lost Packages

Lost packages can have a significant impact on both customers and merchants. For customers, not receiving their package is more than just an inconvenience. It can lead to frustration and disappointment, especially if the item was something they were eagerly anticipating. This negative experience can tarnish their perception of your brand and make them hesitant to shop with you again.

For merchants, lost packages mean financial losses. You lose money on the original transaction and incur additional costs for sending a replacement. Moreover, handling claims and addressing customer complaints takes up valuable time and resources that could be better spent on growing your business.

## Common Reasons for Lost Packages

To effectively manage lost packages, you need to understand the common reasons behind them. Here are some typical scenarios:

Carrier Issues: Mistakes on the carrier's side, such as misrouting, damaged trucks, or mishandling, can result in lost packages.

Incorrect Addresses: Customers sometimes forget to update their addresses when they move. As a result, packages are delivered to their old address and may never reach the intended recipient.

Mis-delivery: Packages might get delivered to the wrong address, where the recipient might not bother to return them.

Theft: Porch piracy is a growing problem, with thieves stealing packages left unattended on doorsteps.

## Strategies for Handling Lost Packages

Proactively managing lost packages is crucial for maintaining a positive customer experience. Here are some strategies to consider:

Clear Communication: Keep your customers informed about their package's status with tracking information and updates. This transparency helps manage their expectations and reduces anxiety about potential delays or losses.

Accurate Data Entry: Encourage customers to double-check their shipping information before completing their purchase. Implement address verification software to catch errors before packages are sent out.

Insurance and Protection: Offer shipping insurance to cover the cost of lost or damaged packages. This can provide peace of mind for both you and your customers.

Quick Resolutions: When a package is reported lost, act swiftly. Investigate the issue, communicate with the carrier, and keep the customer updated. Promptly send a replacement or issue a refund to maintain customer trust.

## Best Practices for Prevention

Preventing lost packages should be a top priority. Here are some best practices to help reduce the risk:

Use Reliable Carriers: Partner with reputable carriers known for their reliability and excellent customer service. This can significantly reduce the chances of lost packages.

Implement Tracking Systems: Use advanced tracking systems that provide real-time updates on package locations. This can help identify issues early and improve the chances of recovering lost packages.

Address Verification: Integrate address verification tools into your checkout process to ensure customers provide accurate shipping information.

Customer Education: Educate your customers about the importance of providing accurate shipping details and keeping their addresses up to date. Include reminders in your order confirmation emails and on your website.

## Dealing with Customer Complaints

Handling customer complaints about lost packages requires empathy and efficiency. Here's how to manage these situations:

Listen and Empathize: Understand that the customer is frustrated and disappointed. Listen to their concerns and empathize with their situation.

Provide Solutions: Offer clear solutions, such as sending a replacement or issuing a refund. Be transparent about the steps you're taking to resolve the issue.

Follow Up: After resolving the issue, follow up with the customer to ensure they're satisfied with the outcome. This extra step shows you care about their experience and can help rebuild trust.

## Leveraging Technology

Technology can play a crucial role in managing lost packages. Here are some tech solutions to consider:

Automated Tracking Systems: Implement automated systems that provide real-time updates and alerts for both you and your customers.

Customer Portals: Create customer portals where customers can easily track their orders, report issues, and communicate with your support team.

Data Analytics: Use data analytics to identify patterns and trends in lost packages. This can help you address underlying issues and improve your shipping processes.

Managing lost packages is a vital part of ensuring a positive post-purchase experience. By understanding the common reasons for lost packages and implementing strategies to prevent and handle them, you can minimize their impact on your business. Clear communication, accurate data entry, and leveraging technology are key components of an effective lost package management strategy. By prioritizing customer satisfaction and addressing issues promptly, you can maintain trust and loyalty, ultimately driving the success of your business.

## Building a Feedback Loop

Once a transaction is complete and the shipment leaves the warehouse, your job isn't over—it's just beginning. You need to stay vigilant about what's happening with the customer's order. Ensure customers receive tracking numbers and can easily see where their package is. Many people, myself included, are obsessed with tracking their orders.

At a minimum, provide tracking information on your website and send it via email. Modern services can also send updates when the package is out for delivery or delayed. It's crucial to inform customers about delays due to weather or other factors. Without updates, customers might worry and initiate chargebacks, which is a bad situation for any business.

When the shipment finally reaches the customer, your work still isn't done. Ask for feedback on their experience. You can request product reviews, service ratings, and feedback on their customer support interactions. This not only helps you improve your service but also provides valuable insights into your products.

For instance, BarkBox once introduced a fun challenge on Instagram, asking customers to post pictures of their dogs ruining their toys. This allowed them to see which parts of the toys were most prone to damage and improve their product design.

## The Importance of Tracking and Updates

Tracking is a crucial part of the post-purchase experience. Customers want to know where their package is at all times. Providing them with a tracking number and regular updates ensures they feel informed and in control. It reduces anxiety and builds trust in your brand.

In my business, we made it a point to offer comprehensive tracking information. Customers could check the status of their orders on our website and receive email notifications at each stage of the delivery process. This proactive approach significantly reduced customer inquiries about order status and increased satisfaction.

## Communicating Delays

No one likes delays, but they do happen. Whether it's due to weather, logistical issues, or other factors, keeping your customers informed is vital. If a package is delayed, notify your customers immediately. Explain the reason for the delay and provide an updated delivery estimate.

We found that transparent communication about delays actually strengthened customer loyalty. When customers know you're being upfront with them, they're more likely to be understanding and patient. This transparency helps prevent frustration and reduces the likelihood of chargebacks or negative reviews.

## Gathering Feedback

Once the customer receives their order, it's time to gather feedback. This step is crucial for understanding how well you're meeting customer expectations and where you can improve. Here are some effective ways to collect feedback:

Email Surveys: Send a follow-up email asking customers to rate their experience and provide any comments. Keep the survey short and straightforward to encourage more responses.

Product Reviews: Encourage customers to leave reviews on your website. These reviews not only provide valuable feedback but also help other potential customers make purchasing decisions.

Customer Support Feedback: After any interaction with customer support, ask customers to rate the service they received. This can help you identify areas where your team might need additional training or resources.

## Using Feedback to Improve

Collecting feedback is only useful if you act on it. Regularly review the feedback you receive and look for patterns or recurring issues. Use this information to make data-driven decisions about how to improve your products and services.

For instance, we noticed several customers mentioned difficulties with our website's navigation. By addressing this issue and making the site more user-friendly, we saw a noticeable increase in customer satisfaction and conversion rates.

## Engaging Customers Through Social Media

Social media is a powerful tool for engaging with your customers and gathering feedback. Encourage your customers to share their experiences on platforms like Instagram, Facebook, and Twitter. Not only does this provide you with valuable insights, but it also helps build a community around your brand.

Take a page from BarkBox's book. Their Instagram challenge, where customers posted pictures of their dogs with BarkBox toys, was a huge success. It provided BarkBox with direct feedback on product durability and design, while also creating a fun and engaging way for customers to interact with the brand.

## Creating a Continuous Improvement Loop

Building a feedback loop isn't a one-time task. It's an ongoing process that requires regular attention and refinement. Here's how to maintain an effective feedback loop:

Regularly Update Surveys: Keep your surveys fresh by periodically updating the questions to cover new products or services.

Act on Feedback: Implement changes based on the feedback you receive. Let customers know when their feedback has led to improvements. This shows them you value their input.

Monitor Social Media: Stay active on social media and monitor what customers are saying about your brand. Engage with them, respond to comments, and address any concerns promptly.

Review Metrics: Regularly review key metrics like customer satisfaction scores, return rates, and product reviews to gauge the impact of your improvements.

## Leveraging Technology for Feedback

Technology can make the feedback process more efficient and effective. Here are some tools you can use:

Survey Platforms: Use platforms like SurveyMonkey or Google Forms to create and distribute surveys easily.

Customer Relationship Management (CRM) Systems: CRM systems can help you track customer interactions and feedback, providing a comprehensive view of customer satisfaction.

Social Media Monitoring Tools: Tools like Hootsuite or Sprout Social can help you monitor social media mentions and engage with customers more effectively.

Building a robust feedback loop is essential for enhancing the post-purchase experience and driving continuous improvement. By providing tracking information, communicating delays, gathering feedback, and leveraging social media, you can ensure your customers feel valued and heard. Acting on the feedback you receive helps you refine your products and services, ultimately leading to increased customer satisfaction and loyalty. Remember, the post-purchase experience is just the beginning of your relationship with your customers—make it count.

## Upsells

Running a regular subscription service opens up numerous opportunities for upsells. Upsells are a great way to increase the lifetime value (LTV) of a customer. By offering additional products or premium services, you can significantly boost your average order value and deepen customer loyalty.

## The Power of Upsells

Upselling is about providing additional value to your customers. It's not just about increasing sales but about enhancing the customer experience. When done right, upsells can make customers feel like they're getting more from their subscription, which can increase satisfaction and retention.

For instance, Dollar Shave Club started by selling razor blades. They then expanded to offer aftershave, shaving creams, and other grooming products. These complementary items were one-time purchases that significantly increased their average order value. Customers appreciated the convenience of getting all their grooming essentials in one place, which made them more likely to stick with the subscription.

## Types of Upsells

There are several effective strategies for upselling:

Complementary Products: Offer products that naturally complement the subscription. For example, if you sell a subscription box for pet owners, you could upsell items like toys, grooming products, or premium treats.

Toss-ins: These are one-time purchases that can be bundled with a subscription box to reduce shipping costs. For example, if someone subscribes to a monthly snack box, offer them a special gourmet snack as a toss-in.

Flash Sales: Run limited-time discounts on specific items. Flash sales create urgency and can boost sales of specific products. For example, you could offer a 24-hour discount on a popular item in your store.

Premium Upcharges: Introduce more expensive items as optional add-ons. For instance, if your customers subscribe to receive scarves, offer an option to add a premium brand scarf for an additional cost.

## Understanding Customer Needs

Successful upselling starts with understanding your customers' needs and preferences. This is where CRM (Customer Relationship Management) marketing comes into play. Use your CRM system to gather insights into your customers' buying behaviors, interests, and preferences.

## Using CRM Marketing

CRM marketing can help you tailor your upsell offers more effectively. Here's how:

Data Collection: Gather data on your customers' purchase history, browsing behavior, and preferences. This information can help you identify which products they're most likely to be interested in.

Segmenting Customers: Divide your customer base into segments based on their behavior and preferences. For example, you might have a segment of customers who frequently purchase grooming products and another segment interested in premium brands.

Personalized Offers: Use the data from your CRM to create personalized upsell offers. For instance, if a customer frequently buys organic snacks, offer them a special deal on a new organic snack product.

Timing: Timing is crucial for upselling. Offer upsells at the right moment, such as during the checkout process or in follow-up emails after a purchase.

## Implementing Upsell Strategies

Implementing upsell strategies requires careful planning and execution. Here's a step-by-step approach to get started:

Identify Upsell Opportunities: Look at your current product offerings and identify items that could be upsell opportunities. Consider products that complement your main subscription offering or premium versions of existing products.

Create Attractive Offers: Design upsell offers that are attractive and provide real value to your customers. This could be a discount on a bundle, a limited-time offer, or a premium product that's not available elsewhere.

Promote Your Upsells: Use multiple channels to promote your upsell offers. This includes your website, email marketing, social media, and during the checkout process. Make sure the offers are visible and easy to add to their purchase.

Track and Analyze: Monitor the performance of your upsell offers. Use analytics to see which offers are most popular and which ones aren't performing as well. This will help you refine your strategy over time.

## Real-World Example: Dollar Shave Club

Dollar Shave Club is a great example of a company that successfully implemented upsells. They started by selling affordable razor blades through a subscription model. Once they built a customer base, they expanded their product line to include aftershave, shaving creams, and other grooming products.

These additional products were marketed as convenient add-ons that customers could easily include with their regular shipments. By understanding their customers' needs and preferences, Dollar Shave Club was able to create upsell offers that were relevant and attractive, significantly increasing their average order value.

Upsells are a powerful way to increase the lifetime value of your customers and boost your revenue. By offering complementary products, toss-ins, flash sales, and premium upcharges, you can enhance the customer experience and provide additional value. Understanding your customers' needs through CRM marketing is key to creating effective upsell offers. Implementing these strategies thoughtfully can lead to increased satisfaction, loyalty, and growth for your subscription business. Remember, upselling is not just about making more sales; it's about making your customers feel valued and catered to.

## Preventing Churn

In the subscription business, churn is a term everyone fears. Churn occurs when you lose a customer, and it comes in two forms: voluntary and involuntary.

### Understanding Voluntary Churn

Voluntary churn happens when customers choose to unsubscribe. They might have enough of your product, or their financial situation might have changed. Regulations now require that customers can easily cancel their subscriptions. However, you can still try to retain them. Offer options to pause or skip their subscription or provide discounts for the next month. This is where creativity and A/B testing come into play.

To combat voluntary churn, it's crucial to understand why customers are leaving. Conduct exit surveys to gather feedback. Are they unhappy with the product quality, or do they feel it's no longer a good value? Use this information to improve your offerings and address any common concerns.

Pause and Skip Options: Give customers the flexibility to pause or skip their subscription if they're not ready for another shipment yet. This keeps them on board without the pressure of monthly deliveries.

Special Offers: Offer discounts or special promotions to incentivize customers to stay. A 10% discount on the next month's subscription might be enough to retain someone considering canceling.

Customer Engagement: Engage with your customers regularly through personalized emails, newsletters, and social media. Keep them excited about your products and informed about any new features or upcoming releases.

Improving Product Value: Continuously enhance the value of your product. Introduce new items, improve existing ones, and listen to customer feedback to make necessary adjustments.

A/B Testing: Test different retention strategies to see what works best. This could include varying the timing of your discount offers or testing different messaging in your communications.

## Addressing Involuntary Churn

Involuntary churn occurs when you can't charge the customer for the following month. This can happen for various reasons, typically categorized as soft or hard declines. Hard declines are cases where the card was closed or reported stolen—situations where you can't recover the charge. Soft declines might be due to insufficient funds or general declines. This is where you need to work with your subscription billing platform to optimize your recovery strategy.

Most billing platforms include basic dunning management, which is the process of retrying failed payments. Some use simple methods, like retrying every two days for five attempts, while others use advanced algorithms that consider various data points. External services like Butter or Vindicia offer sophisticated dunning management solutions. In our case, about 20% of initial payment attempts fail, but we recover around 80% of those failed charges, which significantly impacts our bottom line.

## Steps to Optimize Payment Recovery

Dunning Management: Use advanced dunning management tools to automate the retry process. These tools can help recover a significant portion of failed payments by intelligently scheduling retries.

Communication: Send timely and friendly reminders to customers about their failed payments. Inform them about the issue and provide an easy way to update their payment information.

Flexible Payment Options: Offer multiple payment options to reduce the likelihood of payment failures. Accepting credit cards, PayPal, and other payment methods can help cater to different customer preferences.

Payment Updating: Make it easy for customers to update their payment information. Provide clear instructions and a user-friendly interface for updating expired or invalid card details.

Proactive Monitoring: Monitor your payment recovery process regularly to identify any patterns or issues. This allows you to adjust your strategies and improve your recovery rates continually.

Real-World Example: Reducing Churn

In our business, we faced a significant challenge with involuntary churn. Initially, about 20% of our payment attempts failed. By implementing advanced dunning management solutions and improving our communication with customers, we managed to recover around 80% of those failed charges. This recovery significantly impacted our bottom line, helping us retain more customers and maintain steady revenue growth.

The post-purchase experience is a critical part of running a successful subscription business. From handling lost packages to building a feedback loop, offering upsells, and preventing churn, each step plays a vital role in ensuring customer satisfaction and business growth. By focusing on these aspects, you can create a positive experience that keeps customers coming back and drives your business forward.

Preventing churn, both voluntary and involuntary, is essential for maintaining a healthy subscription business. Use flexible options, special offers, and personalized engagement to reduce voluntary churn. Implement advanced dunning management and proactive communication strategies to address involuntary churn. By taking these steps, you can improve customer retention, enhance satisfaction, and drive long-term growth for your business.

CHAPTER 8

# MASTERING ANALYTICS

In Walter Isaacson's biography of Elon Musk, there's a fascinating story about the Gigafactory in Nevada. For many months, it was a failure because they couldn't achieve their desired production rate of 5,000 cars a week. They couldn't figure out what was wrong until they delved deep into the production line, examining every point of the production cycle and understanding how much time people spent on each operation.

At this point, they introduced weekly meetings to review all the metrics related to their business. Unfortunately, this level of scrutiny doesn't happen in most organizations. Typically, people focus on a few big metrics, like the number of subscribers, revenue, and weekly charges. These are often reviewed at the end of the month, which is too late to make timely changes.

As a business owner, you need to focus on your business metrics regularly. Becoming a data-driven organization requires understanding your business from top to bottom—from customer acquisition to order fulfillment. It also means having the right tools and systems to track all these metrics effectively.

One of the key lessons from Tesla's experience is the importance of regularly reviewing your business metrics. It's not enough to look at them monthly or quarterly. By then, it might be too late to make necessary adjustments. Weekly reviews can help you stay on top of your operations and identify issues before they become significant problems.

When we implemented weekly metric reviews in my business, it transformed our approach. We could quickly see what was working and what wasn't, allowing us to make timely adjustments. This proactive approach helped us improve efficiency and better meet our goals.

Understanding your metrics deeply is crucial. It's not just about knowing the numbers but understanding what they mean and how they impact your business. Tesla's deep dive into their production cycle is a perfect example. They didn't just look at the total production numbers; they examined every step in the process to identify bottlenecks and inefficiencies.

In our business, we applied a similar approach. Instead of just tracking overall sales, we broke it down into different channels, customer segments, and product lines. This detailed analysis revealed insights that we wouldn't have noticed otherwise, allowing us to make more informed decisions.

## Becoming a Data-Driven Organization

To truly leverage the power of analytics, you need to become a data-driven organization. This means integrating data into every aspect of your business and making decisions based on solid evidence rather than gut feelings.

We started by ensuring that all our key processes were tracked and measured. We used tools like Google Analytics for our website, CRM systems for customer interactions, and inventory management software for order fulfillment. By consolidating this data into a central dashboard, we could easily monitor our performance and make data-driven decisions.

Having the right tools and systems in place is essential for effective tracking and analysis. There are many options available, depending on your specific needs and budget. Some popular tools include Google Analytics, Mixpanel, Looker, and Tableau.

For example, we used Google Analytics to track website traffic and user behavior. This helped us understand where our visitors were coming from, what they were looking for, and how we could improve their experience.

Identify Key Metrics: Start by identifying the key metrics that drive your business. These could include sales numbers, customer acquisition costs, conversion rates, and fulfillment times. Make sure these metrics align with your overall business goals.

Set Up Tracking Systems: Implement the necessary tools and systems to track these metrics. This might involve using software like Google Analytics, CRM systems, or custom dashboards.

Schedule Weekly Reviews: Set up a regular schedule for reviewing these metrics. Weekly reviews are ideal, as they allow you to spot trends and issues quickly. Make sure these meetings are a priority and involve key team members.

Analyze and Act: During the reviews, delve into the numbers. Look for trends, anomalies, and areas for improvement. Discuss potential actions and assign responsibilities for implementing changes.

Monitor and Adjust: After making changes, continue to monitor the impact on your metrics. Be prepared to make further adjustments as needed to achieve your goals.

Tesla's meticulous approach to analyzing their production metrics offers valuable lessons for any business. By regularly reviewing detailed metrics and being willing to delve deep into the data, they were able to identify and address issues that were hindering their success.

In our business, we've embraced this approach and seen significant improvements. By focusing on data-driven decision-making, we've been able to optimize our operations, improve customer satisfaction, and drive growth.

Mastering analytics is crucial for the success of any business. By learning from Tesla's approach to production metrics, we can understand the importance of regular metric reviews, deep analysis, and data-driven decision-making. Implementing the right tools and systems, and committing to a rigorous review process, can help you stay ahead of potential issues and make informed decisions that drive your business forward.

Embrace the power of data, and you'll be well-equipped to navigate the complexities of running a successful business.

## Key Subscription Metrics

Subscription businesses are no longer unique, so the metrics are well-known. However, calculating these metrics can be tricky, as there are multiple ways to interpret them. Let's start with the basics and then dive deeper.

The most common metrics include:

- Total number of subscribers
- Revenue
- Number of new subscribers since the last check-in
- Number of charges run last week

I recommend having weekly meetings to review your business performance. Start with these basic metrics and then move to more specific ones like churn. There are two types of churn: voluntary and involuntary, and they should be tracked separately.

Next is LTV (lifetime value), a crucial but tricky metric because it requires future projections. Typically, you'll build projections for 12, 24, and 36 months. Once you have your LTV, the next metric to consider is CAC (cost of acquisition). This is usually split into organic CAC (from organic channels) and paid CAC (from paid channels like Facebook or Google).

The golden standard is the CAC to LTV ratio. This should be at least 3:1, indicating healthy growth with significant organic growth. Besides these metrics, monitor all your marketing campaigns, upsell tactics, and any skip subscriptions.

To effectively run a subscription business, you need to track several key metrics. The first and most obvious is the total number of subscribers. This gives you a snapshot of your customer base and helps you gauge the overall size of your business.

Revenue is another fundamental metric. It tells you how much money you're making, which is crucial for assessing the financial health of your business. Tracking the number of new subscribers since the last check-in helps you understand your growth rate and the effectiveness of your acquisition strategies. Lastly, the number of charges run last week can give you insight into the consistency and reliability of your revenue stream.

## The Importance of Weekly Reviews

Regularly reviewing these metrics is essential. I hold weekly meetings to go over our business performance. This frequent check-in allows us to spot trends early, address issues promptly, and make data-driven decisions. It's not just about looking at numbers; it's about understanding what they mean and how they impact our business.

Churn is a critical metric for subscription businesses. It refers to the number of subscribers who cancel their subscriptions within a given period. There are two types of churn: voluntary and involuntary. Voluntary churn happens when customers choose to cancel their subscriptions. Involuntary churn occurs when subscriptions are canceled due to issues like expired credit cards or failed payments.

Tracking these separately is important because they have different implications and solutions. For example, voluntary churn might indicate a problem with your product or service, while involuntary churn could be addressed by improving your payment processes.

## Calculating Lifetime Value (LTV)

LTV is a crucial metric because it represents the total revenue you can expect from a customer over their entire relationship with your business. Calculating LTV can be tricky because it involves making future projections. Typically, you build projections for 12, 24, and 36 months. This helps you understand how much a customer is worth to your business in the long term.

## Understanding Cost of Acquisition (CAC)

CAC is the cost of acquiring a new customer. It's essential to split this into organic CAC (customers acquired through organic channels like SEO or word of mouth) and paid CAC (customers acquired through paid channels like Facebook or Google ads). This differentiation helps you understand the effectiveness and efficiency of your marketing efforts.

### The CAC to LTV Ratio

The CAC to LTV ratio is a critical metric for assessing the health of your subscription business. Ideally, this ratio should be 3:1, meaning the lifetime value of a customer should be three times the cost of acquiring them. This indicates that your business is growing sustainably and profitably. A lower ratio might suggest that you're spending too much on acquisition relative to the value you're getting from each customer.

### Monitoring Marketing Campaigns and Upsell Tactics

Besides these core metrics, it's important to monitor all your marketing campaigns. Understanding the performance of each campaign helps you allocate your budget more effectively and optimize your marketing strategy.

Upsell tactics are another crucial aspect. Track how well your upsell efforts are working and adjust your approach based on what the data tells you. This can significantly increase the lifetime value of your customers by encouraging them to spend more over time.

### Managing Skip Subscriptions

Finally, keep an eye on skip subscriptions. These are instances where customers choose to skip a delivery or pause their subscription temporarily. While not as serious as churn, frequent skips can indicate dissatisfaction or issues with your product or service. By monitoring this metric, you can address potential problems before they lead to cancellations.

Mastering key subscription metrics is essential for running a successful subscription business. Start with the basics like total subscribers, revenue, new subscribers, and weekly charges. Then, dive deeper into metrics like churn, LTV, and CAC. Regularly review these metrics in weekly meetings to stay on top of your business performance. Understanding and optimizing these metrics will help you make informed decisions, improve customer retention, and drive sustainable growth. By focusing on data-driven strategies, you can navigate the complexities of the subscription model and ensure the long-term success of your business.

## Business Operation Metrics

Tracking subscription metrics is only half the story. Once you acquire a customer, numerous other activities happen across different departments. You need to closely monitor how you run the business.

A significant part of this is operations and order fulfillment. Track the cost of fulfilling one box, the cost of goods for each box, and the specifics of each component. If you offer different types of boxes (e.g., one for the first month and another for subsequent months), track these separately.

Monitor operational details such as injuries in your warehouse (if you run your own) or associated costs from your 3PL partner. For example, know the costs of sending boxes of different sizes and weights. From operations, move to finance—often overlooked because it's seen as boring, but it's the beating heart of your business.

Find a good partner in your finance department or accounting team. Understand GAAP (General Accounting Principles), how revenue is calculated, deferred revenue, and, most critically, cash flow. Cash flow is the one metric that determines whether your business is alive or dead. It encompasses all charges into your bank account and payments out, so you need to manage your invoices correctly to ensure you can always meet payroll and pay vendors.

## Monitoring Operational Metrics

Operations and order fulfillment are the backbone of any subscription business. You need to know exactly how much it costs to fulfill each box. This includes the cost of goods, packaging, shipping, and labor. Tracking these costs allows you to understand your margins and identify areas where you can cut expenses.

When we first started, we meticulously tracked the cost of every component in our boxes. This granular approach helped us identify cost-saving opportunities, like bulk purchasing or finding more affordable suppliers. It's also essential to differentiate between different types of boxes if you offer variations for new and returning customers.

## Warehouse and 3PL Metrics

If you manage your own warehouse, keeping track of operational details like injury rates is crucial. This not only ensures a safe working environment but also helps in controlling costs associated with workplace accidents. If you use a 3PL partner, understand their cost structure thoroughly. Know the costs involved in sending boxes of different sizes and weights, and monitor these expenses closely.

For example, we discovered that shipping costs for certain box sizes were disproportionately high. By tweaking our packaging design, we managed to reduce shipping costs without compromising on the customer experience.

## Importance of Financial Metrics

While often seen as mundane, financial metrics are vital to your business's health. Your finance department or accounting team is crucial in helping you understand and manage these metrics. It's important to grasp GAAP (Generally Accepted Accounting Principles), which dictate how revenue should be calculated and reported. Deferred revenue is another critical concept, especially in subscription businesses where payments are often received in advance.

But the most critical metric of all is cash flow. Cash flow reflects the movement of money in and out of your business. It's essential to manage your invoices and ensure that you have enough liquidity to meet your obligations, such as payroll and vendor payments. Without positive cash flow, your business can quickly run into trouble.

One of the first steps we took was to establish a solid relationship with our finance team. They helped us understand the nuances of revenue recognition and deferred revenue, which are particularly important for subscription businesses. Regularly reviewing our cash flow statements allowed us to stay on top of our financial health and make informed decisions.

Efficiency in operations directly impacts your bottom line. By closely monitoring metrics such as the time taken to fulfill an order, error rates, and return rates, you can identify bottlenecks and areas for improvement.

In our business, we implemented regular audits of our fulfillment process. This involved tracking the time taken from when an order was received to when it was shipped out. Any delays were analyzed to identify root causes, and we implemented changes to streamline the process. This proactive approach helped us reduce order fulfillment times and improve customer satisfaction.

Strong relationships with your suppliers are crucial. Regularly review their performance and costs to ensure you're getting the best value. If possible, negotiate better terms or explore alternative suppliers.

We set up quarterly reviews with our key suppliers to discuss performance and pricing. This not only helped us maintain good relationships but also ensured that we were aware of any potential issues or opportunities for cost savings.

Maintaining high-quality standards is essential for customer retention. Implement a robust quality control process to ensure that every product that leaves your warehouse meets your standards.

In our business, we established a quality control team responsible for inspecting a sample of every batch of products. Any defects or issues were reported immediately, allowing us to address them before they reached the customer. This commitment to quality helped us build trust and loyalty among our customers.

Leveraging technology can significantly enhance your operational efficiency. Invest in software solutions that automate and streamline various aspects of your business, from order processing to inventory management.

We adopted an integrated ERP (Enterprise Resource Planning) system that connected our inventory management, order processing, and accounting functions. This integration provided real-time insights into our operations and allowed us to make data-driven decisions.

Tracking business operation metrics is crucial for the success of any subscription business. From monitoring the cost of goods and order fulfillment to understanding financial metrics like cash flow, these insights help you run your business more efficiently and effectively. Establish strong relationships with your finance team and suppliers, invest in quality control, and leverage technology to streamline your operations. By focusing on these metrics and continuously improving your processes, you can ensure the long-term success and sustainability of your business.

## Implementing Business Intelligence

Once you know which metrics to track, the next step is organizing all this data. Unfortunately, no single tool currently provides all these metrics in one place. Large organizations often use ERP (Enterprise Resource Planning) systems, but these are expensive and don't always offer comprehensive metrics.

Most businesses, including many successful ones, still rely on Excel or Google Sheets for their business intelligence (BI) systems. Here's how to do it: Create a Google Doc for easy sharing within your organization. List all the metrics discussed earlier, and update them every Monday based on the previous week's data.

Hold a meeting every Tuesday to discuss any irregularities or anomalies in the metrics and formulate a plan for the week. For instance, if your CAC is decreasing and your CAC to LTV ratio is healthy, you might decide to increase your marketing spend. Or, if you notice volatility in inventory returns, it might indicate a payment industry issue that needs attention.

Certain trends, like the decrease in inventory returns between February and April due to tax refunds, can impact your metrics positively. Establishing a robust BI system early in your company's life will be crucial for its success.

## Building a BI System with Google Sheets

While sophisticated BI tools and ERP systems are ideal, they can be cost-prohibitive for smaller businesses. Excel and Google Sheets offer a flexible, cost-effective alternative. Here's how to get started:

Set Up a Google Sheet: Create a Google Sheet that can be easily shared with your team. This ensures everyone has access to the most up-to-date data.

List Your Metrics: In the sheet, list all the key metrics you've identified. These should include subscriber count, revenue, new subscribers, charges run, churn rates, LTV, and CAC.

Regular Updates: Update these metrics every Monday with data from the previous week. Consistency is key to maintaining accurate and useful data.

When we first implemented this system, it was a game-changer. We could quickly identify trends, spot issues, and make informed decisions without investing in expensive software.

## Weekly Review Meetings

Holding regular review meetings is essential for staying on top of your metrics. Schedule these meetings for every Tuesday. Here's a simple agenda to follow:

Review Metrics: Go through each metric and note any significant changes or anomalies.

Identify Trends: Look for patterns or trends in the data. For example, a consistent increase in CAC might suggest a need to revise your marketing strategy.

Plan Actions: Based on your analysis, decide on the actions to take. Assign responsibilities and set deadlines to ensure follow-through.

These meetings became a cornerstone of our strategy. They kept everyone aligned and focused on our goals. We could adapt quickly to changes and stay proactive in addressing potential issues.

## Using Metrics to Guide Decisions

Your metrics are not just numbers; they're insights that should guide your business decisions. For example:

Marketing Spend: If your CAC is decreasing and the CAC to LTV ratio is healthy, consider increasing your marketing spend to accelerate growth.

Inventory Management: If you notice a spike in inventory returns, it might indicate issues with product quality or fulfillment. Investigate further to pinpoint the cause and take corrective action.

Seasonal Trends: Understand how seasonal trends impact your business. For instance, tax refunds between February and April might boost sales, leading to a decrease in inventory returns.

In our case, tracking these metrics helped us make more informed decisions about where to allocate resources. We could see which strategies were working and which needed adjustment.

## Addressing Anomalies

Anomalies in your metrics can signal underlying issues. When you spot an anomaly, dig deeper to understand its cause. Here's how:

Investigate: Look into the specific metric showing the anomaly. Is it a one-time issue or part of a larger trend?

Consult Your Team: Discuss the anomaly with your team to get different perspectives. They might have insights you haven't considered.

Develop a Plan: Create a plan to address the issue. This might involve changing a process, reallocating resources, or adjusting your strategy.

For example, we once noticed an unexpected increase in customer churn. By investigating, we discovered it was due to a payment processing issue. Addressing it quickly helped us reduce churn and improve customer satisfaction.

## The Importance of Trend Analysis

Understanding trends is crucial for making strategic decisions. Regularly analyze your metrics to identify both positive and negative trends. This helps you stay proactive rather than reactive.

One trend we noticed was a yearly decrease in inventory returns between February and April, likely due to tax refunds. Recognizing this pattern allowed us to plan better for inventory management and marketing strategies during this period.

## Establishing a Robust BI System

A robust BI system is vital for long-term success. Here's why:

Informed Decision-Making: Accurate, up-to-date data allows you to make informed decisions quickly.

Proactive Management: Regular reviews and trend analysis help you stay ahead of potential issues.

Alignment and Accountability: Weekly meetings ensure everyone is on the same page and accountable for their responsibilities.

Implementing a business intelligence system, even with simple tools like Google Sheets, can significantly enhance your ability to track and analyze key metrics. Regular updates and weekly review meetings keep your team aligned and focused on actionable insights. By leveraging these metrics, you can make informed decisions, address anomalies promptly, and stay ahead of trends. Building this system early in your company's life will set you up for long-term success, ensuring you can adapt and thrive in a competitive landscape.

CHAPTER 9

# RUNNING THE BACK OFFICE

If you've ever been involved in production, you've probably heard about Toyota's system. It was created by the Toyoda family for their factories, and in the software world, it's known as the Kanban system. Toyota invested heavily in their production processes to ensure maximum efficiency. One of their biggest challenges was space. Japan, being an island, has limited space for farming, production, or construction.

Toyota aimed to minimize the number of parts stored and create a seamless production cycle where goods moved smoothly through the conveyor line without unnecessary storage. They developed meticulous processes for each step, ensuring everything worked flawlessly. This approach allowed them to conquer the American market and become a highly efficient organization. The Toyota production system quickly became a global success and a model for others to follow.

Now, let's dive a bit deeper into what makes Toyota's system so effective and how it has influenced my approach to running the back office.

When I first learned about Toyota's production system, I was amazed by its simplicity and effectiveness. The idea of having a just-in-time inventory system, where parts arrive exactly when they are needed, was revolutionary. It cut down on the costs associated with storing large amounts of inventory and reduced waste. This concept of lean manufacturing really resonated with me.

Another key lesson from Toyota's system is the importance of continuous improvement, or "kaizen." Toyota's commitment to constantly refining and improving their processes is something I strive to emulate. In our back office, we regularly review our procedures to identify bottlenecks and inefficiencies. By encouraging a culture of continuous improvement, we not only boost efficiency but also foster innovation and employee engagement.

The idea of standardizing processes is another crucial takeaway. Toyota's meticulous attention to detail in standardizing each step of the production process ensures consistency and quality. Our operations teams created detailed standard operating procedures (SOPs) for various tasks. This not only ensures that everyone follows the same steps but also makes it easier to train new employees and maintain a high level of quality in our work.

One practical application of this in our back office is in our document management system. We've standardized how projects are being run, discussed, and documented, which has streamlined our workflow and reduced errors. By having clear guidelines and templates, we've been able to maintain consistency and improve overall efficiency.

Moreover, Toyota's focus on eliminating waste has been instrumental in shaping our approach. They identified seven types of waste in production: overproduction, waiting, transportation, excess processing, inventory, motion, and defects. By targeting these areas, they were able to create a more efficient production system. We've adopted a similar mindset, constantly looking for ways to eliminate waste in our processes. For example, we've reduced waiting times by improving our communication channels and streamlined our approval processes to avoid unnecessary delays.

The implementation of visual management tools is another aspect of Toyota's system that we've embraced. In their factories, Toyota uses visual cues to communicate information quickly and effectively. Inspired by this, we've incorporated visual management tools in our office, such as dashboards and status boards. These tools provide real-time updates on project statuses, deadlines, and key performance indicators, helping us stay on track and make informed decisions promptly.

Reflecting on my journey of integrating Toyota's principles into our back office operations, I realize that it's not just about adopting a set of techniques. It's about embracing a mindset of efficiency, continuous improvement, and waste reduction. This approach has not only made our operations more effective but has also cultivated a culture of excellence and innovation.

The impact of Toyota's production system extends beyond manufacturing. Its principles can be applied to various aspects of business, including back office operations. By focusing on efficiency, continuous improvement, and waste reduction, we've been able to create a more streamlined and productive work environment. As we continue to refine our processes and adopt new strategies, the lessons from Toyota will remain a guiding force in our pursuit of operational excellence.

## Order Fulfillment

The reason I mention Toyota's production system is that it's something you should emulate in your organization. You might have different departments—marketing, software engineering, etc.—but they all need to work together seamlessly. A smooth back office operation is crucial for customer satisfaction.

Order fulfillment is a critical part of this process. You can either manage your own warehouse or work with a 3PL (third-party logistics) organization. Regardless of your choice, certain things must always be done correctly. First, ensure you have enough product in your warehouse to fulfill orders. If orders pile up and you can't fulfill them, customers will get upset, leading to chargebacks and negative social media feedback.

You must have complete visibility into your inventory. If your inventory levels are too high, it ties up money that could be better used elsewhere. On the flip side, running out of stock can delay order fulfillment. Maintain a strict SLA (service level agreement) around the speed and accuracy of order fulfillment. Orders shouldn't wait, and none should be lost. Use business intelligence dashboards to monitor how long orders stay in the queue, identify any bottlenecks, and ensure timely fulfillment.

In the world of order fulfillment, taking a page from Toyota's book can make all the difference. Just as their production system relies on seamless coordination, your various departments—whether it's marketing, software engineering, or customer service—need to work like a well-oiled machine. The back office operations are the backbone of customer satisfaction, and order fulfillment sits right at the heart of this.

When it comes to fulfilling orders, there are two main routes: managing your own warehouse or partnering with a third-party logistics (3PL) provider. Each option has its own set of challenges and advantages, but no matter which path you choose, some key principles remain constant.

First and foremost, you need to ensure you have enough products in your warehouse to meet demand. If orders start piling up and you can't fulfill them promptly, your customers will be frustrated. This not only leads to chargebacks but also sparks negative feedback on social media. And trust me, nothing damages a brand's reputation faster than unhappy customers voicing their complaints online.

Inventory management is another crucial aspect. You need to have complete visibility into your stock levels. If you overstock, you tie up valuable capital that could be used elsewhere in your business. Conversely, if you run out of stock, you risk delaying orders, which again, leads to customer dissatisfaction. Striking the right balance is key.

Maintaining a strict Service Level Agreement (SLA) for the speed and accuracy of order fulfillment is vital. Orders should never wait in the queue longer than necessary, and losing an order should be out of the question. Using business intelligence dashboards can be a game-changer here. These tools allow you to monitor how long orders stay in the queue, identify any bottlenecks, and ensure that fulfillment happens swiftly and accurately.

In my own experience, adopting these principles has transformed our order fulfillment process. Due to complexity of our business we decided to run our warehouse ourselves, which has allowed us to optimize evert bit of a process. This essentially lead to a better customer experience because order were fulfilled faster, with higher accuracy and care overall.

For instance, we use advanced inventory management software that syncs with our e-commerce platform in real-time. This ensures that our stock levels are always accurate, and we're never caught off guard by a sudden spike in demand. We've set up automated alerts to notify us when stock levels are low, so we can reorder before running out.

Additionally, our business intelligence dashboard has been a crucial tool in optimizing our order fulfillment process. By providing real-time data on order status, fulfillment times, and potential bottlenecks, we can quickly identify and address any issues that arise. This has significantly reduced the time orders spend in the queue and improved our overall efficiency.

Another aspect we've focused on is the accuracy of order picking and packing. Mistakes in this area can lead to returns, which are costly and time-consuming. We've implemented a double-check system where two team members verify each order before it's shipped. This extra step has drastically reduced errors and increased customer satisfaction.

Communication is also key. We've set up automated email notifications to keep customers informed at every stage of the order fulfillment process. From the moment they place an order to the time it arrives at their doorstep, customers receive updates on the status of their shipment. This transparency has built trust and reduced the number of inquiries our customer service team receives.

Reflecting on these improvements, I realize that the lessons from Toyota's production system have been invaluable. By focusing on efficiency, visibility, and continuous improvement, we've created a streamlined order fulfillment process that meets our customers' expectations and supports our business growth.

Whether you choose to manage your own warehouse or work with a 3PL provider, the principles of efficient order fulfillment remain the same. Ensure you have enough product to meet demand, maintain visibility into your inventory, and uphold strict SLAs for speed and accuracy. Use technology to your advantage, and never stop looking for ways to improve. By doing so, you'll not only satisfy your customers but also position your business for long-term success.

## Shipping and Logistics

The COVID-19 pandemic taught us how fragile the global supply chain can be. There are several aspects to consider here. First, how do you get your goods delivered to your warehouse? The most cost-efficient way is usually by sea for international production or by truck for local production. However, you need to understand the production cycle and factors that affect it.

For example, if you're purchasing from China, be aware of the Chinese New Year, during which factories shut down for the entire month of February. You need to order enough stock in December to last until May or April. Sometimes, you might need to expedite shipping by air, but this is costly and usually reserved for urgent needs like promotional campaigns or photo shoots.

Next is local logistics—getting goods to your customers. In the US, there are many carriers. Some handle the first mile, picking up goods, while others handle the last mile, delivering to the customer. USPS often handles the final delivery. Understanding how these logistics work helps you provide a great customer experience while keeping costs efficient.

When the COVID-19 pandemic hit, it really exposed just how fragile our global supply chain is. Suddenly, everything we took for granted about shipping and logistics was turned on its head. So, how do you make sure your goods get delivered to your warehouse? The cheapest way is usually by sea for international production or by truck for local production. But, it's not as simple as it sounds. You really need to understand the production cycle and the factors that affect it.

Aside from international shipping there's the issue of local logistics—getting goods from your warehouse to your customers. In the US, there are a ton of carriers. Some handle the first mile, which is picking up goods from your warehouse, while others handle the last mile, delivering to the customer. Often, USPS will take care of the final delivery. Understanding how these logistics work is crucial to providing a great customer experience while keeping costs down.

In my own experience, adapting to these logistics challenges has been a learning curve. We've had to be proactive about our ordering schedules, especially around major holidays and global events that could disrupt supply chains. For instance, during the pandemic, we started placing larger orders earlier to avoid potential delays. This required us to carefully manage our inventory levels and cash flow to ensure we weren't overstocked but also didn't run out of essential items.

We've also had to explore different shipping options. While sea freight remains the most cost-effective for large shipments, we've occasionally had to rely on air freight for time-sensitive deliveries. This was particularly true during the height of the pandemic when sea routes were heavily congested. Although more expensive, air freight ensured that our high-priority items reached us in time to meet customer demands.

Another critical aspect of shipping and logistics is understanding the intricacies of local delivery networks. In the US, the logistics landscape is quite complex, with various carriers specializing in different parts of the delivery process. Some carriers are great for long-distance hauling, while others excel in urban deliveries. We've partnered with multiple carriers to optimize our delivery times and costs. For example, we use a combination of DHL for the initial stages of delivery and rely on USPS for the final leg. This hybrid approach has helped us maintain timely deliveries even during peak seasons.

To streamline our logistics further, we've invested in technology. Real-time tracking systems allow us to monitor our shipments throughout the entire supply chain. This visibility helps us anticipate delays and communicate effectively with our customers. We've also implemented automated systems for inventory management, which sync with our shipping schedules to ensure we always have the right stock levels. These systems alert us when inventory is running low, so we can reorder before we run out.

One of the biggest lessons we've learned is the importance of flexibility. The pandemic showed us that no matter how well-planned your logistics are, unforeseen events can and will happen. Being able to pivot quickly and find alternative solutions is key to keeping your operations running smoothly. For example, when one of our primary carriers faced significant delays, we quickly shifted some of our shipments to another carrier that wasn't as affected. This agility minimized the impact on our delivery times and kept our customers satisfied.

Customer communication has also been a game-changer. Keeping customers informed about the status of their orders helps manage their expectations and builds trust. We've set up automated notifications that provide updates at every stage of the delivery process. From the moment an order is placed to the time it arrives at the customer's doorstep, they receive timely updates.

This transparency has significantly reduced customer inquiries and increased satisfaction.

Reflecting on these experiences, I realize that shipping and logistics are more than just getting products from point A to point B. It's about creating a reliable, efficient system that can adapt to challenges and still deliver exceptional service. By understanding the complexities of both international and local logistics, planning ahead, investing in technology, and maintaining flexibility, we've been able to navigate the uncertainties of the global supply chain and keep our customers happy.

Whether you're dealing with international shipping or local deliveries, the principles remain the same: plan ahead, stay informed, invest in technology, and be ready to adapt. By following these guidelines, you can create a resilient logistics system that not only meets but exceeds customer expectations, even in the face of global disruptions.

## WMS, OMS, and IMS

In running a back office, various systems help keep things organized. Three key systems are the Warehouse Management System (WMS), Order Management System (OMS), and Inventory Management System (IMS).

### IMS: Inventory Management System

The IMS provides visibility into your inventory. It tracks how much inventory you have on hand, what's being packed, and what's expected to arrive. Plus, it helps manage the costs associated with inventory. Knowing exactly what you have and where it is can prevent overstocking or running out of products. This system is crucial because it ensures that you always have the right amount of stock to meet customer demand without tying up too much capital in unsold goods.

In my own experience, implementing an effective IMS has been a game-changer. Before we had a good system in place, we often found ourselves either overstocked or out of crucial items. This led to unnecessary expenses and sometimes unhappy customers. Now, with a robust IMS, we can forecast demand more accurately and reorder products just in time, which has streamlined our operations and improved cash flow.

### OMS: Order Management System

The OMS is all about managing orders. Often, the final product customers receive isn't the same as what you initially received. For example, a subscription box might include packing materials, promotional items, and the main products. The OMS configures and designs the final list of items in the shipment, creates shipping labels, and organizes order batches to optimize fulfillment costs.

From my perspective, an OMS is indispensable for handling complex orders. We deal with a variety of products that need to be assembled before shipping. The OMS ensures everything comes together perfectly. It creates the shipping labels, arranges the items in the most efficient way, and keeps everything on track. This has not only cut down on errors but also sped up our fulfillment process. Customers get their orders quicker, and we save on shipping costs.

### WMS: Warehouse Management System

The WMS is the backbone of any warehouse operation, whether you run it yourself or use a 3PL provider. It tracks all operations, including how many orders are being processed, the costs, the number of people involved, and the current shipment status. It's the operational nerve center of your business.

Using a WMS has transformed our warehouse operations. Before we had one, keeping track of everything was a logistical nightmare. Now, we can see exactly what's happening at any given moment. The WMS shows us how many orders are being processed, who's working on what, and where every item is. This visibility has allowed us to optimize our workflows, reduce labor costs, and improve accuracy.

In addition to these systems, integrating them has been a major focus. When the IMS, OMS, and WMS work together seamlessly, it creates a cohesive and efficient operation. For instance, when an order comes in, the OMS coordinates with the IMS to check inventory levels and then with the WMS to initiate the picking and packing process. This integration ensures that orders are fulfilled accurately and promptly.

Reflecting on our journey, I remember the days before we had these systems in place. Back then, everything was manual, which meant a lot of room for error and inefficiency. Orders would sometimes get lost, inventory counts were off, and fulfilling complex orders was a headache. Implementing these systems didn't happen overnight. It required careful planning, investment, and training. But the payoff has been immense. We've seen significant improvements in our efficiency, accuracy, and customer satisfaction.

One of the most significant benefits of these systems is the data they provide. We now have access to real-time data on our operations, which helps us make informed decisions. For example, we can analyze order patterns to predict busy seasons and prepare accordingly. We can also track key performance indicators (KPIs) like order accuracy, fulfillment speed, and inventory turnover. This data-driven approach has enabled us to continuously improve our processes and stay ahead of the competition.

Moreover, these systems have helped us scale our business. As we've grown, managing our back office operations has become more complex. But with the right systems in place, we've been able to handle increased order volumes without compromising on quality or speed. The scalability of these systems means that as we continue to grow, they can grow with us, supporting our operations every step of the way.

Looking forward, I see even more potential for optimizing our back office operations. Advances in technology, such as artificial intelligence and machine learning, are opening up new possibilities. For example, AI can help predict demand more accurately, optimize inventory levels, and even automate some aspects of order fulfillment. We're excited to explore these innovations and see how they can further enhance our efficiency and customer satisfaction.

WMS, OMS, and IMS are essential tools for running an efficient back office. They provide the visibility, coordination, and data needed to manage inventory, fulfill orders, and optimize warehouse operations. By implementing and integrating these systems, we've transformed our operations, improved our customer experience, and positioned ourselves for continued growth. As we look to the future, we're committed to leveraging technology to stay at the forefront of our industry and deliver the best possible service to our customers.

CHAPTER 10

# THE CUSTOMER EXPERIENCE

E-commerce and subscription services are about more than just the products we sell. It's about the overall experience and service we provide to our customers. Customer experience is critical because if things go wrong or people have questions, they need to get answers and feel taken care of. When customers have a positive experience, they appreciate it and recommend your service to others.

A great example of exceptional customer service is Chewy. Traditionally, call centers measure success by how many calls an agent can handle in a day. This means agents often rush through calls. Chewy, however, took a different approach. They told their customer service agents to spend as much time as needed to ensure the customer is happy. They started tracking customer satisfaction and how often customers returned to the website after a call.

One story that stands out involves a woman who called Chewy to return a bag of dog food because her dog had died. Instead of processing the return, Chewy suggested she donate the food to a local dog shelter. This simple act of kindness quickly went viral on social media, driving a lot of positive attention and goodwill towards Chewy. This is why Chewy stands out as an outstanding company.

Customer experience is everything. In the age of social media and instant reviews, a single negative experience can quickly become public, affecting your brand's reputation. On the flip side, a positive experience can turn customers into loyal advocates who recommend your service to others. It's not just about selling a product; it's about building relationships and trust with your customers.

When we started Scentbird we wanted our customers to feel valued and heard. We set up multiple channels for customer support, including email, phone, and live chat, ensuring that help was always available. We also trained our customer service team to handle inquiries with empathy and patience, taking a page out of Chewy's book.

What sets Chewy apart is their commitment to going above and beyond for their customers. They understand that customer loyalty is built on trust and positive interactions. By encouraging their agents to spend as much time as needed with each customer, they ensure that every interaction is meaningful and leaves a lasting impression.

For example, I remember an instance where a customer reached out to us about a delayed order. Instead of just apologizing and providing a new delivery date, our team took it a step further. We offered a small discount on their next purchase and sent a handwritten note thanking them for their patience. The customer was thrilled and shared their positive experience on social media, bringing us more attention and goodwill.

Chewy's approach to customer service isn't just about solving problems; it's about building emotional connections. The story of the woman returning dog food after her pet's death is a perfect example. By suggesting she donate the food to a shelter, Chewy showed empathy and understanding, turning a potentially negative experience into a positive one.

At Scentbird, we strive to create these kinds of emotional connections. We keep detailed records of our customers' preferences and purchase history, allowing us to personalize our interactions. For instance, if a customer frequently buys a specific product, we might reach out with a special offer or a personalized recommendation. These small gestures help build a deeper connection with our customers.

Traditional call centers measure success by the number of calls handled or the speed of resolution. Chewy flipped this model on its head by focusing on customer satisfaction and loyalty. They track how often customers return to their site after an interaction and use this data to continually improve their service.

We've adopted a similar approach in our business. Instead of focusing solely on metrics like call volume or response time, we prioritize customer feedback and satisfaction. We regularly survey our customers to understand their needs and pain points, using this information to make informed improvements to our service.

Chewy's success story teaches us that exceptional customer service isn't about following a script or meeting quotas. It's about understanding your customers, treating them with respect, and always striving to exceed their expectations. This philosophy has guided our approach to customer service and helped us build a loyal customer base.

Implementing Chewy's strategies in your own business can have a profound impact. Start by empowering your customer service team to take the time needed with each customer. Train them to listen actively and respond with empathy. Create a system for tracking customer satisfaction and use this data to drive continuous improvement.

Additionally, consider how you can create emotional connections with your customers. Personalize your interactions and go the extra mile to show that you care. Whether it's a handwritten note, a small token of appreciation, or simply taking the time to listen, these efforts can make a significant difference.

Mastering customer experience is about more than just resolving issues—it's about creating positive, memorable interactions that build loyalty and trust. By taking inspiration from Chewy's approach, we can improve our customer service and build stronger relationships with our customers. Remember, a happy customer is not only a repeat customer but also a powerful advocate for your brand. Embrace the philosophy of going above and beyond, and you'll see the benefits in both customer satisfaction and business growth.

## Enhancing On-Site Customer Experience

There are multiple ways customers can communicate with you: phone calls, chatbots, social media, and emails. However, it all starts with your website. Your UX (user experience) design must ensure that people understand how your website works, where to click, and how to find and review products. Conduct A/B tests and usability tests to refine this experience continually.

Additionally, make sure your website complies with the ADA (American Disability Act) or similar regulations in your country.

This ensures that people with disabilities can use your site. Companies can help you achieve this compliance.

A user-friendly website is the foundation of a great customer experience. The design should be intuitive, guiding users effortlessly through their journey. Every element, from the navigation bar to the product pages, should be designed with the user in mind.

When I was developing my website, I focused on simplicity and clarity. I made sure that users could easily find what they were looking for without feeling overwhelmed. Clear calls to action, organized menus, and a straightforward checkout process were essential.

To ensure your website meets user expectations, conduct regular A/B tests and usability tests. A/B testing involves comparing two versions of a webpage to see which one performs better. Usability tests involve observing real users as they navigate your site to identify any pain points or areas of confusion.

For example, when we were unsure whether to place the "Add to Cart" button at the top or bottom of the product page, we conducted an A/B test. We found that placing it at the top resulted in more clicks and higher conversions. Similarly, usability tests helped us identify issues with our search function, leading to improvements that made it easier for customers to find products.

An FAQ section is a must-have for any website. It addresses common questions and concerns, providing customers with quick and easy access to the information they need. However, don't stop there. Embedding relevant FAQ snippets into specific pages, such as payment and shipping pages, can provide immediate answers and reduce the need for customers to navigate away from their current page.

For instance, if a customer is on the checkout page and has a question about shipping times, having a snippet right there can save them the trouble of searching through the entire FAQ section. This improves their experience and reduces the likelihood of cart abandonment.

Providing clear and accessible contact options is essential. Make sure your contact information, including email addresses and chat options, is easy to find. A visible "Contact Us" link in the header or footer can make a big difference.

Customer feedback is invaluable. Always close the feedback loop by addressing customer questions and concerns. If a customer reaches out with an issue or a question, respond promptly and make sure they feel heard. This not only resolves their immediate problem but also builds trust and loyalty.

For example, if a customer reports a problem with your website, thank them for bringing it to your attention and let them know when it has been fixed. Follow-up emails or messages can reinforce that you value their input and are committed to providing a great experience.

## Reducing Abandonment

One of the biggest challenges in e-commerce is reducing abandonment. When customers don't understand something or encounter an issue, they're likely to leave your site and choose a competitor. To prevent this, make sure your website is easy to navigate, your contact options are clear, and your FAQ section is comprehensive.

Additionally, use tools like exit-intent popups to engage customers who are about to leave your site. Offer them a discount or ask if they need assistance. These small interventions can significantly reduce abandonment rates.

Enhancing on-site customer experience is about making your website intuitive, accessible, and responsive to customer needs. Regular testing, accessibility compliance, comprehensive FAQs, and easy contact options are all critical components. By focusing on these areas, you can create a seamless and enjoyable experience for your customers, leading to increased satisfaction and loyalty. Remember, a happy customer is more likely to return and recommend your service to others, driving growth and success for your business.

## Handling Customer Experience Issues

Handling significant problems that arise in your business is crucial. These issues could be a data breach, supplier issues affecting shipments, or problems with your third-party logistics (3PL) provider. The first step is to acknowledge the problem. Communicate openly on your website, letting customers know about the issue and what you're doing to resolve it.

For example, we once faced a situation where a snowstorm in Georgia prevented our 3PL company from getting employees to the warehouse. We communicated this to our customers, who appreciated that we prioritized the safety of our partners' employees. Transparency fosters trust and support from your customers.

Next, manage customer expectations by establishing a clear communication channel. This could be through your website, social media, or email. Provide regular updates about the status of the problem and your progress in resolving it. Share your action plan and any changes that occur.

Respond to customer questions with empathy and positivity. Yes, people will be upset and angry, but handling their concerns well ensures the continued success of your business.

## Acknowledging the Problem

The first step in handling any major issue is acknowledging it. Ignoring problems or trying to downplay them can erode customer trust. When something goes wrong, be upfront about it. Clearly state what the problem is and how it impacts your customers.

For example, during the snowstorm in Georgia, our 3PL provider couldn't get their employees to the warehouse. We immediately posted an update on our website and social media channels, explaining the situation. We assured our customers that we were working on solutions and that the safety of our partners' employees was our top priority.

## Transparency and Communication

Transparency is key in maintaining customer trust during a crisis. Openly sharing information about the problem and your efforts to resolve it shows that you value your customers and their experience.

We used multiple communication channels to keep our customers informed. Our website had a dedicated section for updates, our social media accounts provided real-time information, and we sent out regular emails to keep everyone in the loop. This approach ensured that no customer was left in the dark.

## Managing Expectations

Setting and managing customer expectations is critical. Let your customers know what to expect in terms of delays, solutions, and timelines. Being clear about what you can and cannot do helps prevent frustration and confusion.

During the snowstorm incident, we informed our customers about potential delays in their shipments. We provided estimated timelines for when the issue would be resolved and assured them that we were doing everything possible to expedite their orders once conditions improved.

## Providing Regular Updates

Regular updates are essential in keeping customers informed and reassured. Even if there are no significant changes, letting customers know that you're still working on the issue is important.

We committed to providing daily updates on the situation. These updates included what steps we were taking, any progress made, and any changes to the expected resolution time. This consistent communication helped our customers feel connected and informed throughout the process.

## Responding with Empathy

When customers reach out with questions or complaints, respond with empathy. Understand that they're frustrated and may feel inconvenienced. Acknowledge their feelings and show that you care about resolving their issues.

We received many messages from concerned customers during the snowstorm. Our customer service team was trained to respond with empathy and positivity. We listened to their concerns, provided clear answers, and reassured them that we were doing everything possible to resolve the situation. This approach helped diffuse anger and built goodwill.

## Learning from the Experience

After resolving the issue, take time to reflect and learn from the experience. What went well? What could have been handled better? Use these insights to improve your processes and prepare for future challenges.

Following the snowstorm incident, we conducted a thorough review of our response. We identified areas where we could improve our communication and logistics processes. These lessons were invaluable in helping us handle future issues more effectively.

## Implementing Preventative Measures

To minimize the impact of similar issues in the future, implement preventative measures. This could involve diversifying your supplier base, improving your logistics planning, or enhancing your data security measures.

For instance, after the snowstorm, we explored options for diversifying our 3PL partners to ensure we had backups in place. We also improved our logistics planning to better handle weather-related disruptions. These steps helped us become more resilient and better prepared for future challenges.

## Building Customer Loyalty

Handling issues effectively can actually strengthen customer loyalty. When customers see that you care about their experience and are willing to go the extra mile to resolve problems, they're more likely to remain loyal to your brand.

We received numerous messages of appreciation from customers who were impressed by our transparency and dedication during the snowstorm incident. Many of them became even more loyal to our brand, knowing that we prioritize their satisfaction.

## Leveraging Systems and Automation

We live in an era where AI and automation are integral to customer support. While I believe in maintaining human touchpoints to connect with real customers and understand their needs, automation plays a vital role. Start with a unified system for handling all customer requests, such as Zendesk or Gladly. These ticketing systems can handle emails, social media interactions, text messages, phone calls, and communications through platforms like WhatsApp or Telegram.

Automate common requests, such as “Where is my order?” or “Cancel my subscription,” using AI tools like Digital Genius or Zoom Virtual Agent. These tools can integrate with platforms like Shopify to streamline responses. For more complex requests, your team can handle them, using the feedback to improve your services continuously.

Lastly, address chargebacks effectively. Chargebacks are a reality, but tools can automate the process by connecting with your payment provider and compiling necessary documentation for the issuer bank. This saves you time and money.

## The Role of AI and Automation

Incorporating AI and automation into customer support isn’t about replacing human interaction but enhancing it. Automation can handle repetitive and straightforward tasks, allowing your team to focus on more complex issues that require a personal touch.

For instance, when I first started using automation tools, I was amazed at how much time it saved. Tasks like tracking orders and managing cancellations were handled efficiently by AI, freeing up our customer service agents to address more pressing concerns. This not only improved our response times but also increased customer satisfaction.

For example, a frequent question we get is, "Where is my order?" By automating this request, customers receive instant updates on their order status without needing to wait for a human response. This not only improves the customer experience but also reduces the workload on your team.

While automation can manage routine inquiries, complex issues still require a human touch. However, AI tools can assist here too. They can triage requests, directing simpler issues to automated responses and escalating more complicated ones to your team. This ensures that your agents can focus on the cases that truly need their expertise.

Our team uses automation to gather initial information about complex issues, which helps streamline the process when a human agent takes over. This hybrid approach allows us to maintain high efficiency while providing personalized service where it matters most.

One of the benefits of integrating AI and automation is the continuous feedback loop. By analyzing the types of requests and issues that arise, you can identify areas for improvement in your products or services. This data-driven approach helps you make informed decisions to enhance customer satisfaction.

For instance, after implementing automated responses, we noticed a recurring issue with one of our products. Using this insight, we were able to address the problem proactively, reducing the number of related customer inquiries and improving overall satisfaction.

## Managing Chargebacks

Chargebacks can be a significant headache for any business. Automating the chargeback process can save you time and reduce financial losses. Tools that connect with your payment provider and compile necessary documentation for the issuing bank can streamline this process.

We integrated a chargeback management tool that automated the collection of evidence and communication with banks. This not only sped up the resolution process but also increased our success rate in disputing chargebacks. By reducing the manual effort involved, we saved valuable time and resources.

CHAPTER 11

# EMBRACE THE UNEXPECTED

There's a saying I love: "We do things not because they are simple, but because we thought they were simple." When you start your entrepreneurial journey, you often don't know what you're getting into. That's what makes it so exciting. When we started Scentbird back in 2014, I had no idea about the challenges and learning curves I would face. Each year brings something new and fascinating, though the challenges can sometimes be tough.

It's crucial to have support. Ideally, you're not running your business solo but with a team of co-founders who bring different skills and support each other through challenges. Friends and family also play a vital role. They're always there to support and understand your dreams. However, it's important to keep work problems separate from personal relationships. If necessary, don't hesitate to seek help from a therapist.

Starting Scentbird was like jumping into the deep end of the pool without knowing how to swim. I quickly learned that entrepreneurship isn't a straight path. There are twists, turns, and unexpected detours. But these surprises are part of what makes the journey so rewarding. Each challenge I faced taught me something new and helped me grow both personally and professionally.

One of the first unexpected lessons was the sheer amount of resilience required. There were days when everything seemed to go wrong. Suppliers would delay shipments, marketing campaigns would flop, or technical issues would crash our site. During those times, it felt like the world was against us. But I learned to pick myself up, analyze what went wrong, and find a solution. This resilience became a cornerstone of my entrepreneurial mindset.

Another unexpected lesson was the importance of adaptability. When we first launched Scentbird, we had a clear vision of what we wanted to achieve. However, as we grew, we had to adapt to market demands, customer feedback, and industry trends. Flexibility allowed us to pivot when necessary and stay relevant in a competitive market. I realized that being too rigid could stifle growth and limit potential.

Support, as I mentioned earlier, was another critical factor. My co-founders and I leaned on each other during tough times. Each of us brought unique strengths to the table, which helped us tackle problems from different angles. This collaboration was invaluable. We celebrated our successes together and supported each other through setbacks. It wasn't just about sharing the workload; it was about creating a supportive environment where everyone felt valued.

Friends and family also played a significant role in my journey. Their encouragement and belief in my vision kept me going, especially during the tough times. However, I quickly learned the importance of setting boundaries. Mixing work stress with personal relationships can create tension. By keeping work issues separate from personal time, I maintained a healthy balance and ensured that my relationships remained strong.

Another key aspect of embracing the unexpected was learning from failures. Every setback was an opportunity to learn and improve. For example, when our initial marketing strategy didn't yield the expected results, we analyzed what went wrong and adjusted our approach. This willingness to learn from mistakes and adapt was crucial for our growth and success.

One of the most significant lessons I learned was the power of perseverance. There were moments when giving up seemed like the easiest option. But each time I pushed through the difficulties, I emerged stronger and more determined. Perseverance taught me that success often comes to those who are willing to keep going, even when the going gets tough.

Embracing the unexpected is a fundamental part of the entrepreneurial journey. It involves being resilient, adaptable, and open to learning from every experience, whether good or bad. Support from co-founders, friends, family, and even therapists is invaluable. Maintaining a healthy work-life balance and taking care of your mental well-being are also essential. By facing challenges head-on and learning from them, you can navigate the unpredictable path of entrepreneurship and emerge successful and fulfilled.

## Preparing for Scale and Managing Cash Flow

One aspect that can make or break an e-commerce or subscription business is cash flow. We've touched on this before when discussing business intelligence, but it's worth emphasizing. Many founders struggle because they don't fully understand cash flow or can't predict their financial situation for the coming months.

When you start business for the first time, managing cash flow could be a huge learning curve. It's one of those things you can't ignore, no matter how exciting your product or service is. In the early days of your business, create a simple Excel spreadsheet to track your cash flow. This FP&A (financial planning and analysis) report should cover the next three to six months. Understand your monthly costs and revenue to build a projection for the future. It's critical to ensure you can always pay your employees. This is one of the most crucial aspects of being an entrepreneur. Whether you use a basic Excel sheet or a sophisticated ERP system, proper cash flow management is essential.

Cash flow management isn't just about tracking money coming in and going out. It's about understanding the timing of those transactions and ensuring you have enough liquidity to cover your expenses, even when sales are slow. Many businesses fail because they run out of cash, not because they're unprofitable. That's why it's vital to get a handle on your cash flow from the very beginning.

Starting with a simple Excel spreadsheet is a practical approach.

Here's how you can set it up:

List All Income Sources: Include all expected revenue, such as sales, subscriptions, and any other income streams. Break it down by month to see a clear picture of your inflows.

List All Expenses: Include everything—salaries, rent, utilities, marketing expenses, and any other costs. Again, break these down by month.

Calculate Net Cash Flow: Subtract your total expenses from your total income for each month. This will give you an idea of your net cash flow.

Forecast for the Future: Use your historical data to project your future cash flow. Look at trends and consider any upcoming expenses or changes in revenue.

This basic setup will give you a clear view of your financial health and help you make informed decisions.

## The Role of FP&A in Scaling

As your business grows, your financial planning and analysis will become more complex. You might need to invest in more sophisticated tools or hire a financial expert to help you manage your finances. At my company, we eventually moved from a simple Excel spreadsheet to an ERP system that integrated all our financial data. This allowed us to manage our cash flow more effectively and make better strategic decisions.

Scaling a business requires significant investment. Whether you're expanding your product line, entering new markets, or increasing your marketing efforts, you'll need capital. Proper cash flow management ensures you have the funds available when you need them.

## Managing Unexpected Expenses

No matter how well you plan, unexpected expenses will arise. Equipment might break down, a marketing campaign might cost more than anticipated, or you might face sudden changes in the market. Having a cash reserve for these situations is crucial. This reserve acts as a financial cushion, giving you the flexibility to handle surprises without jeopardizing your business.

## Understanding Your Break-Even Point

Another important aspect of cash flow management is understanding your break-even point. This is the point at which your total revenue equals your total expenses. Knowing your break-even point helps you set realistic sales targets and pricing strategies. It also gives you a clear understanding of how changes in your costs or revenue impact your profitability.

## The Impact of Payment Terms

The terms you negotiate with your suppliers and customers can significantly impact your cash flow. For instance, if your suppliers require payment within 30 days, but your customers have 60 days to pay you, you could face a cash flow gap. Negotiating better payment terms with suppliers or offering discounts for early payments from customers can help bridge this gap.

## Leveraging Technology for Cash Flow Management

Technology can be a game-changer in managing cash flow. Various software tools can automate your financial tracking, provide real-time insights, and help you forecast future cash flow more accurately. At my company, adopting an ERP system integrated our financial data, making it easier to manage our cash flow and scale our operations.

## Continuous Monitoring and Adjustment

Cash flow management isn't a one-time task. It requires continuous monitoring and adjustment. Regularly review your cash flow statements, compare them against your projections, and adjust your spending and revenue strategies as needed. Staying proactive will help you avoid cash flow problems and keep your business on a steady growth path.

Managing cash flow is a critical component of running a successful business. From creating simple spreadsheets to using sophisticated

ERP systems, the goal is to have a clear view of your financial health and ensure you have the liquidity to meet your obligations. By understanding your cash flow, preparing for unexpected expenses, and leveraging technology, you can scale your business effectively and sustainably.

As you grow, remember that cash flow management is an ongoing process. Stay vigilant, keep learning, and don't hesitate to seek help when needed. With proper cash flow management, you can navigate the challenges of entrepreneurship and build a thriving business.

## The Entrepreneurial Mindset

Entrepreneurship is a unique state of mind. It's something that can't be fully taught, yet there's so much to learn from others. If you're passionate about being an entrepreneur, follow that passion. The highs and lows are intense, but from my experience, it's absolutely worth it.

Finding a mentor is invaluable. Look for someone with a similar business or someone who can teach you essential lessons. It could be in business operations, finance, or any specific area. Don't be shy about reaching out. Use LinkedIn to connect with potential mentors. If you write to ten people, at least one will likely respond. There are countless stories of successful people who weren't afraid to ask for help—Steven Spielberg in movies or Steve Jobs in tech, who secured his first job at Atari after reaching out to the HP director for spare parts.

Learning is a continuous process. As an entrepreneur, you'll need to understand finance, operations, production, online design, digital marketing, and more. Dedicate time to learning from others, reading books, attending conferences, and talking to industry experts. This ongoing learning will help you minimize mistakes and grow your business efficiently.

Let me delve a bit deeper into these key aspects.

## Passion and Perseverance

The entrepreneurial journey is filled with ups and downs. It's not a smooth ride, but if you're truly passionate about your idea, that passion will fuel your perseverance. I remember countless nights working late, facing obstacles that seemed insurmountable. But my belief in what I was doing kept me going. Passion isn't just about loving what you do; it's about having the determination to push through the tough times.

## The Value of Mentorship

When I started out, I didn't fully appreciate the value of having a mentor. I thought I could figure everything out on my own. However, the reality is that having someone who has walked the path before you can be incredibly beneficial. They can provide insights, share their experiences, and help you avoid common pitfalls.

I once reached out to a successful entrepreneur in my industry through LinkedIn. I was nervous about whether they would respond, but to my surprise, they did. That connection turned into a valuable mentorship that guided me through some critical decisions. The lesson here is simple: don't hesitate to reach out. Most people are willing to help if you ask.

## Continuous Learning

The world of entrepreneurship is constantly evolving. To stay ahead, you need to be a lifelong learner. This doesn't mean you have to go back to school, but you should always be looking for opportunities to learn. Whether it's reading the latest industry books, attending workshops, or following thought leaders online, there's always something new to discover.

I made it a habit to read at least one book a month related to business or personal development. Books like "The Lean Startup" by Eric Ries and "Good to Great" by Jim Collins provided me with new perspectives and strategies that I could apply to my business. Additionally, attending industry conferences allowed me to network with other entrepreneurs and learn from their experiences.

## Building a Diverse Skill Set

As an entrepreneur, you wear many hats. You'll need to understand a bit of everything, from finance to marketing to product development. This can be daunting, but it's also an opportunity to grow. When I first started, I knew very little about digital marketing. But by taking online courses and experimenting with different strategies, I gradually became more proficient.

It's important to recognize your weaknesses and work on improving them. However, don't be afraid to delegate tasks to experts when necessary. You can't do everything yourself, and hiring skilled professionals can free up your time to focus on what you do best.

## Networking and Relationships

Building a strong network is crucial for any entrepreneur. Relationships with other business owners, industry experts, and potential customers can open doors and provide support when you need it most. Attend networking events, join industry groups, and actively engage on platforms like LinkedIn.

Failure is an inevitable part of entrepreneurship. Not every idea will succeed, and not every strategy will work. The key is to learn from your failures and use them as stepping stones to success. When one of my marketing campaigns didn't yield the expected results, I didn't see it as a setback. Instead, I analyzed what went wrong, adjusted my approach, and tried again.

Embracing failure means not being afraid to take risks. Calculated risks can lead to significant rewards, but you must be prepared for the possibility of failure. The important thing is to keep moving forward and not let setbacks discourage you.

## Balancing Work and Life

Entrepreneurship can be all-consuming, but it's essential to maintain a balance between work and personal life. Burnout is real, and it can hinder your productivity and creativity. Make time for yourself, your family, and your friends. Engage in activities that recharge you, whether it's exercising, hobbies, or simply relaxing.

I found that scheduling personal time in my calendar helped me maintain this balance. Treating personal time with the same importance as business meetings ensured that I didn't neglect my well-being.

## Wisdom from Rockefeller

One of the resources I highly recommend is " The 38 Letters from J.D. Rockefeller to his son." This collection offers timeless advice. Each letter is unique and can be read in any order. Rockefeller shares insights on business, partnerships, and life.

One quote that resonates with me is, "I believe that every right implies a responsibility; every opportunity, an obligation; every possession, a duty." Even though these lessons come from the past, they remain incredibly relevant. Learning from historical leaders can provide valuable guidance on your entrepreneurial journey.

## The Value of Timeless Advice

The wisdom found in "Rockefeller Letters to His Son" is remarkable because it transcends time. Although these letters were written in the early 20th century, the principles Rockefeller discussed are still applicable today. His emphasis on responsibility, opportunity, and duty is something every entrepreneur can learn from.

For example, I often found myself overwhelmed by the many decisions I had to make. Reading Rockefeller's letters helped me understand the importance of staying grounded and maintaining a sense of duty towards my team and customers. His advice reminded me that with every opportunity I pursued, there was an inherent responsibility to act ethically and thoughtfully.

Rockefeller's quote about responsibility and duty really struck a chord with me. He believed that "every right implies a responsibility; every opportunity, an obligation; every possession, a duty." This mindset is crucial for any entrepreneur. It reminds us that our actions have consequences, and we must act with integrity and responsibility.

Another significant theme in Rockefeller's letters is mentorship. He often advised his son on various aspects of business and life, acting as a mentor. This reinforces the idea that having a mentor can be incredibly beneficial.

When I sought out mentors, I found that their guidance helped me navigate complex business landscapes. They provided insights that I might not have considered and shared their experiences, which helped me avoid common pitfalls. Mentorship has been a cornerstone of my growth as an entrepreneur, just as it was for Rockefeller's son.

Rockefeller's letters also highlight the importance of continuous learning. He believed in constantly improving oneself and staying informed. This principle is especially relevant in today's fast-paced business environment.

I make it a point to continuously learn and adapt. Whether it's through reading books, attending seminars, or simply staying curious, this habit has helped me stay ahead in my industry. Embracing lifelong learning ensures that you remain competitive and innovative.

Rockefeller often wrote about the balance between business and personal life. He stressed that while business is important, it shouldn't come at the expense of personal well-being and relationships. This advice is crucial for maintaining a healthy work-life balance.

In my own life, I've found that setting boundaries and making time for family and personal interests helps me stay grounded. It prevents burnout and keeps me motivated. Balancing business and life is essential for long-term success and happiness.

Embracing these lessons has helped me grow not only as a business leader but also as an individual. I encourage you to seek out such timeless wisdom and apply it to your own path. The journey of entrepreneurship is challenging, but with the right mindset and guidance, it's incredibly rewarding.

# ABOUT THE AUTHOR

Andrei Rebrov is the co-founder and CTO of ScentBird.com, a leading subscription-based fragrance service with over 600,000 active subscribers. His journey from a novice in web development and the fragrance industry to a successful entrepreneur is both inspiring and instructive. Eleven years ago, Andrei had no experience in creating websites or distinguishing between designer and niche fragrances. Today, he leads a team of 40 talented engineers, driving innovation and excellence at Scentbird.com.

Andrei's and his co-founders' entrepreneurial path began with a simple idea: reducing the risk for people buying perfume online. This idea evolved into helping customers build personalized fragrance collections. Achieving product-market fit was a critical milestone, taking about a year of hard work and perseverance. Andrei's dedication to learning, from coding basics to building warehouse management system, laid the foundation for Scentbird's success.

Made in the USA
Monee, IL
13 September 2026

62980091R00099